The Theory of
Suspension Bridges

The Theory of Suspension Bridges

BY

SIR ALFRED PUGSLEY
O.B.E., D.Sc., F.R.S.

Professor of Civil Engineering in the University of Bristol

LONDON

EDWARD ARNOLD (PUBLISHERS) LTD.

Made in Great Britain at the Pitman Press, Bath

Preface to First Edition

WHATEVER the scientific, professional and educational aims embodied in this book, it must be confessed at the outset that interest in the subject matter is the real cause of its appearance. Opportunities as a schoolboy on holiday to wonder at Telford's Menai Bridge no doubt initiated this interest; the much later challenge of trying, when in the shadow of Brunel's Clifton Bridge, to explain to undergraduates the behaviour of suspension bridges has both intensified and widened it. And in between, difficulties in understanding the scanty and detached references to the subject in modern English textbooks have kept it alive.

It is this lack of an English book on the subject that must be the author's main excuse for this publication. English textbooks on the theory of structures usually either ignore suspension bridge problems or confine their remarks to the parabolic cable and some modified version of Rankine's theory. Comparable American books, until recently, present only a brief version of the elastic theory and do little to link this, on the one hand, with the earlier Rankine theory or, on the other, with the modern deflection theory. One aim of this book, then, is to establish all these theories in their proper historical setting and to show their scientific relationships.

Another, and practically a more important one, is to present a text on the whole matter that is not beyond an honours civil engineering student in his final year and yet is of value also to postgraduate research workers and to professional engineers with special interests in the subject. To do this it has seemed desirable, beyond a natural concentration on the principal theories already mentioned, to open with a more thorough study than usual of the behaviour of a single cable under load and to include some reference to at least the basic ideas of some of the approximate methods of analysis developed in the last few decades.

No writer on suspension bridges can, since 1940, fail to be influenced by the Tacoma Narrows Bridge disaster. An endeavour has here been made, it is believed for the first time in a text such as this, to discuss in outline the natural modes of bridge oscillations and the nature, so far as is at present known, of their excitation by the wind.

This book is the result, directly or indirectly, of the encouragement and help of many friends. To Sir Richard Southwell and Professor A. J. S. Pippard for their influence over the years, to Dr. R. A. Frazer for his kindness, as always, in giving me the freedom of his laboratories, to C. D. Crosthwaite and O. A. Kerensky, of Messrs. Freeman,

v

Fox & Partners, for their helpful criticism of some of my suspension bridge work; to all I render grateful thanks. I must not fail also to mention the generous interest of workers abroad on the theory of suspension bridges, of D. B. Steinman and F. Bleich, of the U.S.A., of A. Selberg, of Norway, and of Professor J. Satterly, of Canada. To Dr. Steinman, of course, I am also, as must be any other modern writer on the subject, specially indebted for the help afforded by his well-known *Practical Treatise on Suspension Bridges*. In seeking to acknowledge all such help I must not suggest that anyone but myself is responsible for the present book, with its shortcomings and, I fear, its errors. I can only hope that these are not too gross in kind or number.

In conclusion, I must express my indebtedness to my secretary, Miss Anne Cornford, for her careful preparation of the typescript, and Dr. P. S. Bulson for his checking thereof. I must also thank my publishers for all they have done for this book; in particular, Mr. F. P. Dunn for the encouragement and help he has given me throughout its preparation.

A. G. P.

Bristol
March, 1956

Preface to Second Edition

WHILE the call for a second edition of this book is gratifying to the author, the main cause is not to be found in any merits of the book itself but in the successful return of British engineers to the construction of suspension bridges. After over one hundred years of almost complete inaction, the decade since the first publication of this book has seen, in rapid succession, the building of the Tamar Bridge, the Forth Bridge, and the Severn Bridge. This has naturally awakened the interest of many civil and structural engineers, and has at the same time, because of the new design features incorporated in some of these bridges, placed British engineers once again in the forefront among suspension bridge designers.

In preparing this edition the author has made a number of alterations and additions without, he hopes, changing the essential nature of the book. The alterations range from the correction of some misprints and other errors kindly pointed out by readers to the rewriting of a number of paragraphs. The aim in each such case has been both to improve the text and to bring the matter involved into line with the results of recent research and with current design thinking.

The same aims account largely for the more substantial additions. These comprise two new chapters—one on suspension bridge towers and the other on the fatigue of suspension bridges—and some new sections on the bearing of digital computers on the application of suspension bridge theories, on natural frequencies of vibration, and on aerodynamic oscillations.

In this way reference has been made to a number of important results of recent research work, though comprehensiveness, even in relation to British work, could not be achieved or sought. The author is much indebted to the authors of this research work and also to two larger publications since the first edition appeared: the English translation of the book on *Dynamic Stability* (*Automobiles, Aircraft, Suspension Bridges*) by Y. Rocard of France, and the two volumes of the Symposium on *Wind Effects on Building and Structures* held at the National Physical Laboratory in 1965.

I must once again express my indebtedness to the publishers of the book.

Bristol
July, 1967

A. G. P.

Contents

CHAPTER 1

Historical Introduction

1.1. It is probable that, just as the modern girder bridge was first suggested in nature by the falling of trees across streams, the suspension bridge owes its origin to the extravagancies of ropes of creepers, vines and other trailing plants in warm countries. As a kind of bridge used by primitive man, it is indigenous in South East Asia, South America and Equatorial Africa, and was probably in use in these regions long before the earliest known contemporary record, which relates to a rope suspension bridge over the Indus River, near Swat, and was written *circa* A.D. 400.*

The native bridges in Assam, built generally between trees as towers, are constructed of bamboo and comprise one or two main cables from which a footway of transverse canes is suspended by more or less vertical rods. The footway is commonly provided with handrail cables. Similar bridges occur in the Himalayas, though these, as elsewhere in South-east Asia, such as Burma, are sometimes built of twisted osier or vine cables. F. W. Robins[1] gives some excellent photographs of some of these primitive bridges, one of which, in Java, is highly modern in form. Indeed, the most primitive arrangement of suspension bridge is probably now most common, curiously enough, in Europe. It consists of two cables, one above the other; the lower is used to walk on and the upper to steady oneself whilst so doing. Though now the cables are of wire, examples of this sort of bridge exist in Sweden and in the Scottish Highlands. There is one in Pantmawr, Montgomeryshire.

1.2. In South America, suspension bridges were early in use among the Incas. The cables were built of aloe or of twisted osiers, the towers were of natural rock, and anchorages were provided by attaching the cables to heavy timber cross beams held fast by rocks. Such bridges had to be overhauled—and the cables replaced—every few years, and local villages were responsible for this maintenance work.

Creepers of various kinds are used by the natives of Equatorial Africa for their suspension bridges, which are often little more

---

* See *Geographical Journal*, August, 1942. Navier, in his *Mémoire sur les ponts suspendus* (1823), describes an iron chain bridge over the Pan-Po River in China known as the Bridge of Iron and traditionally believed to have been erected by a Chinese General in A.D 65.

than assemblies of ropes between tree tops, though primitive anchorages by stakes in the ground sometimes occur.

1.3. The translation of these elementary suspension bridges built of natural ropes into terms of metal occurred first in China. The ropes, there and in some early Tibetan bridges, were replaced by iron chains with links of about one inch diameter bar; and the towers were sometimes of masonry. A fine example of such a bridge, and still existing, is the 200 ft span chain bridge over the Hwa Kiang River. This bridge, built about 1632, has sixteen iron chains, and carries the caravan traffic using the old Imperial highway of the Ming dynasty. Many smaller chain bridges exist in Northern China, and for safety caravan leaders are well accustomed to limit the number of animals crossing a bridge at a given time.

1.4. Western peoples only became interested in suspension bridges on any appreciable scale with the introduction of wrought iron, first, as in China, in the form of chains. In England, wrought iron chains were first forged on a large scale for use as anchor chains for ships, and suspension bridges using such chains tended to arise near the early shipyards. It is believed that the first chain bridge in England was erected over the Tees near Middleton in 1741. It was of primitive character, just a footbridge 2 ft wide, known as the Winch Bridge, and collapsed in 1802. A more ambitious bridge, and one more widely known, was that built over the Tweed by Samuel Brown, a captain in the Navy and a Thames chain maker, in 1820. In this case eye-bars were used for the chains for the first time. The bridge was known as Union Bridge and had a span of 449 ft with a 12 ft roadway for carriages. It was blown down during a storm some six months after its completion. A lesser known but slightly earlier chain suspension bridge was that by Sir John Rennie and Adam Smith over the Humber at Hookstow. It had a span of 130 ft with stone towers, and was built in 1807, but its ultimate fate is unknown. Most of these early chain bridges suffered oscillations in high winds and some collapsed as a result.

The credit for the first iron suspension bridge in the West is usually accorded to James Finley, who built his first chain bridge of 70 ft span across Jacob's Creek, Pennsylvania, U.S.A., in 1796. Two iron chains with specially elongated links were used, and the roadway was 13 ft wide. His later bridge, the Merrimac Bridge, in Massachusetts, U.S.A., built in 1810, is still in existence, though its wooden towers have been replaced by reinforced concrete ones and its chains by wire cables.

1.5. We have here arrived at the stage when the first books on suspension bridges appeared. Navier toured England in 1821 for the purpose of studying its suspension bridges, and must have seen most of those just mentioned. His book, *Mémoires sur les ponts suspendus*, naturally took the form of a report thereon and was published in Paris in 1823. There was great interest in the subject in France at the time. M. Sequin had just proposed to bridge the Rhone near Tournon (this was in fact done in 1825, using wire cables), and he followed Navier's book by one of his own on *Des Ponts en Fil de Fer* in 1824.

We have arrived, too, at the stage when Telford's Menai Bridge (completed in 1826) created a standard of engineering for the future as well as a record in bridge spans throughout the world that remained unbroken till 1834. The Menai Bridge impressed Navier, as it has done most bridge engineers since. Its main span of 580 ft was carried by chains of wrought iron links in plate form with forged eye ends; and from these hung a deck 28 ft in total width, arranged as two roadways. Oscillations of the bridge deck and chains have arisen during storms and some damage has thus occurred from time to time, but the bridge survived with only minor modifications till 1939, when a considerable part of the suspended span was rebuilt. The towers, an outstanding feature of the design, are of masonry of a quality and character that was itself a substantial engineering advance.

1.6. During this first quarter of the nineteenth century, theoretical work on the behaviour of suspension bridges began to develop. Telford consulted the then President of the Royal Society, David Gilbert, on the form of the curve adopted by suspension chains, and Gilbert was thereby led to develop the theory of chains whose sectional area so varied as to ensure uniform stress throughout the span. Away in Russia, where the bridging of the Neva at St. Petersburg (Leningrad) by a suspension bridge was under discussion, Fuss in 1794 developed the theory of the parabolic cable supporting a load that was uniform along the span. It is remarkable that the parabolic form of a cable under such conditions, though previously suspected, was not earlier proved. By the end of the same period, Brunel, working on the design of the Clifton Bridge, made calculations* showing cognisance of all three of the known shapes—simple catenary, catenary of uniform strength (Gilbert's) and the parabola. In none of the designs of this period, however, does there appear any theoretical appreciation of the stiffening function of the light longitudinal girders then prevalent. The problem was the form and design of the cable member; the decking was just a beam-like structure hung therefrom.

* See his original calculation books, now in the possession of the Library of the University of Bristol.

1.7. The second quarter of the nineteenth century saw, in England, many and varied experiments directed towards stiffening suspension bridges. James Dredge, one of the competitors for the design of the Clifton Bridge, sought to do this by using suspension rods that were inclined instead of vertical, the inclination being always towards the span centre; at the same time, and as he believed consistent with this, he tapered the section of the cable by reducing it towards the span centre.[2] His first substantial bridge on this principle was built in wrought iron over the Avon at Bath in 1836; it still survives as a footbridge and has a span of 150 feet. Dredge conducted many model experiments to justify his scheme and built a number of small bridges to this pattern.

At about the same time, a rival constructor of small suspension bridges, Thomas Motley, was advocating and using what he called the *inverted bracket* method of stiffening. His scheme was to replace the ordinary vertical suspension rods by a number of inclined rods radiating from the top of each tower, thereby simulating in tension the systems of compression members in timber that were currently used to radiate from the piers of timber viaducts, and so form timber *brackets*. Motley built the Twerton Bridge, of 120 ft centre span, over the Avon near Bath on this principle in 1838. Contemporary reports emphasised its unusual stiffness for a suspension bridge, and it started a fashion for diagonal suspension rods that later appeared in the Lambeth Bridge by P. Barlow (1861) and the Albert Bridge by R. M. Ordish (1873) over the Thames. Ordish had found that for such longer spans—the Albert Bridge has a central span of 450 ft—the long inclined rods out to the span centre themselves sagged under their own weight and necessitated the reintroduction of a catenary cable over the whole span from which to support, by light vertical rods, the main diagonal rods from the towers.

Thus ended a period of suspension bridge stiffening for spans of from 100 ft to 500 ft by means of inclined suspension rods of various forms. The final products have proved capable of carrying heavy road traffic with little deflection, and the theory of such systems, still without much subsequent improvement, was crystallised in elementary form in Rankine's books on Applied Mechanics (1858) and Civil Engineering (1863).

1.8. By the second half of the century, however, a new approach was arising. Whereas Robert Stephenson, when designing the Britannia Tubular Bridge (1845), had argued that stiffening by means of deck girders would necessitate such large girders that the cables and suspension system might be dispensed with, Peter Barlow now showed[3] by a series of model experiments that much weaker girders would, in fact, be sufficient to spread local loads on to the cables and so avoid large deflections. It was as a result of this that Rankine produced his approximate theory for two and three hinged stiffening girders that has been used so much ever since. By assuming that any concentrated load was spread by the girder uniformly across the whole span on to the cables, he produced the first rational theory of the interaction of cable and girder. At the time, of course, the bridges concerned were of spans of only a few hundred feet, and the stiffening girder was being advocated as an alternative to the inclined rods of Motley and others. It was thus natural for John A. Roebling, in America, when he built his first long span railway bridge, that of 821 ft span over the Niagara Falls, to combine all three methods—the normal cables with vertical rods, the inclined rods radiating from the towers, and the deep heavy stiffening girder; and he did so with such success that the first locomotive (weighing 23 tons) to pass over the bridge caused a central deflection of only $3\frac{1}{2}$ in.[4]

1.9. It was at this stage that the importance and influence of the bridge weight itself upon the bridge stiffness was first realised. Prior to the building of the Niagara Falls Bridge, Roebling had said in a letter[5] to the bridge company concerned:

Although the question of applying the principle of suspension to railroad bridges has been disposed of in the negative by Mr. Robert Stephenson. . . . Any span with fifteen hundred feet, with the usual deflection, can be made perfectly safe for the support of railroad trains as well as common travel.

The success of his 821 ft span railway bridge was reported by him to the same company in 1855, and in doing so he remarks:

Weight is a most essential condition, where stiffness is a great object,

at the same time giving a prophetic warning that stiffness of this sort alone would not prevent oscillation troubles due to wind.

Roebling, from his Niagara experience, had evidently gained an intuitive understanding of the stiffness of a heavy suspension bridge of long span due to gravity forces. This, and his practical development of the construction of wire cables for suspension bridges, led Roebling finally to the successful design and—in 1883, after his death—the successful completion of the great Brooklyn Bridge of 1596 ft span. The stiffening girders of the Niagara Bridge, between its upper and lower decks, were 16 ft deep—i.e. about one-fiftieth of its span; the main trusses of the Brooklyn Bridge are only 17 ft deep—i.e. about one-ninetieth of its span. Roebling still clung to the radiating inclined suspension rods of his earlier designs, but was clearly becoming more confident of the cable gravity stiffness associated with long spans. Brooklyn Bridge was a triumph of intuitive engineering and was heralded as the eighth wonder of the world. It raised such enthusiasm in America that the suspension bridge leapt to the fore for all their new long span bridges, and American engineers became the world experts in this kind.

1.10. With the completion of the Brooklyn Bridge there came also two major steps forward in the theory of suspension bridges— the growth of the "elastic" and "deflection" theories. It seems wrong to attribute the first of these to any one man; it grew rather out of the work of a number spread over some thirty years from 1880 onwards. In 1880 itself, Celeste Clericetti[6] discussed the approximate theory of the Brooklyn type of structure—cable, diagonal and vertical rods and stiffening girder—and was followed in more modern fashion, in 1881, by C. B. Bender[7] on the combined action of an elastic beam suspended from an elastic cable. By 1886, in a paper by Maurice Levy,[8] the essential ingredients of the approximate elastic theory were crystallised. The idea of deciding upon the generally assumed uniform action of the cable on the girder, when the latter is loaded, not arbitrarily as was done by Rankine, but by choosing its intensity so that the vertical deflections of cable and girder matched, was now established.

By this time the parallel theory of arches had developed considerably and Castigliano's[9] strain energy work, with its application to arches, had been published. Many continental engineers, following Navier's early lead, were already accustomed to parallel arch and suspension bridge theory, and it was natural therefore that later developments were cast in strain energy and arch form. Melan's paper[10] of 1906 was translated by Steinman in 1913, and in due course the *elastic theory* of today was standardised by the latter in his own book.[11]

1.11. Alongside this development has come the more accurate *deflection theory*. The fact that the response of a heavy suspension cable, without any stiffening girder, under a concentrated load is non-linear was well known in the first half of the nineteenth century. The mathematical expression of this non-linearity, however, did not appear in general form till the publication of an approximate analysis by an unknown author[12] in 1862; and it was not until 1888 that the first non-linear theory of suspension bridges was evolved by J. Melan.[10] In this theory, which was further developed by Melan in the later version of his work published in 1906, the treatment follows the arch-like form of the elastic theory, but makes allowance for the non-linear behaviour of the chain by recognising its change of shape under a concentrated load and the corresponding change of the tension in the chain. The first practical application of this deflection theory was in the computations for the Manhattan Bridge by L. S. Moisseiff; these were checked by the same method by F. E. Turneaure in 1909, when the bridge, which had a span of

1470 ft, and had been designed by Gustav Lindenthal, was opened. This experience, together with the translation of Melan's work by D. B. Steinman in the same year and published in 1913,[3] effected the very substantial advance in the understanding of suspension bridges that formed a background to the great expansion of their use in America during the next few decades of this century.

1.12. Lindenthal's Manhattan Bridge was followed by Steinman's 1114 ft span Florianopolis Bridge, by the 1750 ft span Delaware River Bridge designed by R. Modjeski and others in 1926, by the beautiful 1200 ft span Mount Hope Bridge of Robinson and Steinman in 1929, and some others of like dimensions. In 1931 came the George Washington Bridge, a further great step forward in size and, by its successful use as a major roadway without stiffening girders, a clear proof of the fact, originally sensed by Navier and later acted upon intuitively by Roebling, that with size and weight comes a natural stiffness. This bridge over the Hudson River at one bound raised the maximum bridge span in the world from 1850 ft to 3500 ft and provided an eight lane roadway with the strength adequate for the addition of a lower deck with four more traffic lanes if required. It was designed by O. H. Ammann and set a standard for a new series of great bridges in America, culminating in the Golden Gate Bridge of J. B. Strauss, with its record span of 4200 ft, in 1937.

In the design of all these great bridges, the *deflection theory* as developed by Moisseiff and Steinman was used, and it was natural that several advances in technique should arise. The best known and most influential has been that due to S. Timoshenko, who in a series of papers[14] showed how the basic differential equation of the theory could be solved by the use of Fourier trigonometric series, and thus provided a more rapid process than was previously available for the calculation of the deflections and bending moments in stiffening girders.

1.13. In England, where the art of suspension bridge construction largely lapsed during the last quarter of the nineteenth century, these American developments in the current century brought about a revival of interest, and in 1939 Southwell[15] showed how the differential equation of the deflection theory could be treated by his Relaxation Process, then under rapid development. At the same time he drew attention to the fact that by his process allowance could be made, for the first time, for the horizontal actions introduced by any displacements, from the vertical, of the suspension rods. Following upon this work, and arising out of the preliminary design calculations for the proposed 3300 ft span Severn Bridge,

C. D. Crosthwaite,[16] of Messrs. Freeman, Fox & Partners, showed how the Relaxation Method could be applied in practice.

The interest engendered by this work, coupled with the conduct of certain full scale experimental work on the Clifton Bridge, has led Pugsley to suggest another method of analysis. H. Bleich[17] had already shown that linearisation of Melan's treatment was possible without undue error and led to considerable computative simplification. Accepting this for the large span bridges of today, in which the cable plays the major structural part, Pugsley[18] suggested the treatment of a suspension bridge in terms of two structures, one the cable and the other the stiffening girder, by matching their deflections, both vertical and horizontal (if necessary), by finite tables of influence or flexibility coefficients. By so doing, variations in the sections or other properties of either cable or girder could be treated, provided only that the behaviour of each was effectively linear. Sufficient has been done with the method to show that linking of the two systems at some ten stations only across the span will give reasonable accuracy, corresponding to the solution of ten simultaneous equations.

Both these methods have been further developed in the past decade, mainly to facilitate their numerical treatment by digital computer. The relaxation method has been used for the design of both the Forth and Severn bridges, and a procedure of the flexibility coefficient kind was employed for the analysis of the stresses in these bridges during erection.

1.14. The foregoing paragraphs relate chiefly to the progressive development of more and more accurate theories, with only passing references to their simplification. But practical design has always called forth various approximate processes, if only as a basis for preliminary designs to be checked by more accurate methods. This is not the place to enlarge on such methods, but reference may be made to a few of the better known or more interesting procedures.

Rankine's method was, of course, itself so simple as to be a direct design method without modification. With the development of the elastic method, however, greater complexity was introduced which Steinman simplified by the diagrams and tables in his standard book.[11] The real difficulty came with the use of the *deflection theory*, and here Steinman has sought to provide an approximate initial treatment on the basis of his design calculation experience. Starting with a non-dimensional stiffness factor relating relevant cable and girder properties, he and a student, A. H. Baker, have built up a system of coefficients, specified by graphs in the above mentioned book, by which to modify the results of the elastic theory.

An alternative approach to this design problem was to follow

Rankine's own basic considerations on the modes of suspension cable deflections, as was in part done by Sir George Airy[19] at the time of a discussion on the Clifton Bridge just after its completion. This approach is essentially that followed in modern times by Hardesty and Wessman, whose 1939 procedure is now the most popular of preliminary design methods.[20]

A further procedure has been tentatively suggested more recently. Arising from his work on flexibility coefficients and on the gravity stiffness[21] of suspended cables generally, Pugsley has noticed that the stiffening girder can be treated approximately as an elastic beam on a linear foundation provided by the cable, and that the stiffness of the latter is approximately constant across most of the span. As a result, approximate information about the bending moments and deflections of the girder due to localised applied loads can be readily calculated.[22] Ways of improving this foundation analogy either by providing guidance for the choice of the effective foundation stiffness[23] or, with more accuracy, by an energy treatment of the flexibility coefficient method[24] have since been developed.

1.15. The gradual development of suspension bridge theory outlined in this chapter has, of course, led to the construction of progressively more economical, more slender and more ambitious structures; and early warnings of the failure of suspension bridges by oscillations in high winds culminated in the major disaster that befell Moisseiff's Tacoma Narrows Bridge in 1940. This was a slender bridge of 2800 ft span that early showed a marked tendency to oscillate, both in flexure and torsion, in the wind, and finally, after a life of only a few months, collapsed as a result of excessive oscillations in a transverse wind of only about 40 m.p.h.

This disaster so shocked the engineering world that major efforts have since been made to understand the nature of this bridge *flutter* problem and to learn how to counter it. Large scale model experiments have been made in both America and England, mainly by Farquharson[25] and Frazer,[26] and several tentative theories, of which perhaps the most advanced is that of F. Bleich,[27] have been put forward. As a result, it has been found that the gravest oscillations can be largely prevented by proper aerodynamic measures applied to the deck and girders of the bridge, and approximate notions[28,29,69] regarding the basis of such measures and other structural needs are available to designers.

Such measures were first adopted in this country in the design of the Tamar and Forth bridges. In both instances, open lattice stiffening girders were adopted and deck structures of high torsional stiffness were achieved by wind bracing at top and bottom of these girders. In the case of the Severn Bridge, the deck structure became a shallow plated box with a cross-section chosen to minimise the formation of eddies in a lateral wind. The measures[30] taken for ensuring its stability have, as a result, become much like those regularly adopted for aeroplane wings.

The construction during the past decade of these three bridges, the Tamar to the design of Messrs. Mott, Hay and Anderson, and the Forth and Severn primarily to designs by Messrs. Freeman, Fox and Partners, has brought British designers back into the forefront of suspension bridge design, much as they were in the days of Telford and Brunel.

CHAPTER 2

The Simple Cable

2.1. A single flexible cable suspended between two fixed points is the simplest possible kind of suspension bridge. The initial problem in such a case is to determine the form adopted by the cable when loaded solely by its own weight, and to find the tension in the cable at any point along its length. The solution of this problem provides a starting point for the consideration of the effects upon a suspended cable of extraneous applied forces, such as arise from the live loads on a practical suspension bridge. This chapter is devoted to the initial problem; the effects of applied loads other than the weight of the cable itself are studied in Chapter 3.

2.2. The Common Catenary

The curve in which a frictionless uniform chain or a perfectly flexible uniform cable hangs when freely suspended between two fixed points is called a catenary. By *perfectly flexible*, we mean that

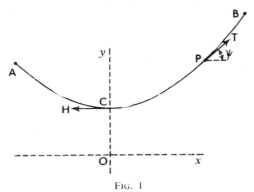

Fig. 1

the cable is incapable of carrying any load save by means of tension directed along its length; the immediate connotation of "uniform" is that the weight per unit length w' of the cable is constant. This defines the classical problem of the common catenary which was first solved by geometers, led by James Bernouilli, in 1691; the earliest English solution was published by David Gregory in 1697.*

---

* E. J. Routh, in Chapter X of his *Treatise on Analytical Statics*, Vol. I, 2nd edition, 1896, gives an interesting footnote on the history of the solution of the catenary or "chainette" problem.

Consider the cable suspended between A and B as shown in Fig. 1, and let C be its lowest point. Let O, the origin for the ordinates x and y, be vertically below C, and let s be the length of the arc measured from C to any point P along the cable. Let the tension in the cable be H at C and T at P, where its inclination, like that of the cable, is ψ.

Treating the cable as inextensible, the weight of the portion CP will be $w's$, and this force, acting vertically through the centre of gravity of the arc CP, will be equilibrated by the tensions H and T. Resolving these forces horizontally and vertically, we have the equilibrium equations

$$T \cos \psi = H, \tag{1}$$

$$T \sin \psi = w's, \tag{2}$$

whence, by writing $H = w'c$ and dividing (2) by (1), we have

$$s = c \tan \psi. \tag{3}$$

This is the intrinsic equation of the catenary, and the constant c is known as the parameter of the catenary.

The Cartesian form of equation (3) may be deduced by writing (3) as

$$c \cdot \frac{dy}{dx} = s,$$

and differentiating to obtain

$$c \cdot \frac{d^2y}{dx^2} = \frac{ds}{dx} = \left\{ 1 + \left(\frac{dy}{dx} \right)^2 \right\}^{\frac{1}{2}}.$$

This gives on integration

$$c \sinh^{-1} \frac{dy}{dx} = x + A,$$

where A is a constant of integration. But as our origin is vertically below the lowest point C, we note that at $x = 0$, $dy/dx = 0$, so that $A = 0$ and

$$\frac{dy}{dx} = \sinh \frac{x}{c}. \tag{4}$$

Integrating again we obtain

$$y = c \cosh \frac{x}{c} + B,$$

where B is another constant of integration. If we make the ordinate OC equal to c, then at $x = 0$, $y = c$, so that $B = 0$, and we have

$$y = c \cosh \frac{x}{c} \tag{5}$$

as the Cartesian equation of the catenary. This gives the shape of the curve adopted by the cable. When required, the length of any arc of the cable is, from (3) and (4), given by

$$s = c \sinh \frac{x}{c}. \tag{6}$$

As for the tension in the cable, by squaring (1) and (2) and adding the results, we have

$$T^2 = w'^2(s^2 + c^2),$$

whence from (5) and (6) we deduce that

$$T^2 = w'^2 y^2$$

and

$$T = w'y. \tag{7}$$

We may therefore note that

(a) the horizontal component H of T is constant and equal to $w'c$,
(b) the vertical component of T at any point P is equal to $w's$,
(c) the resultant tension T at any point P is equal to $w'y$, where y is measured from the directrix Ox.

All the above results depend for their usefulness upon a knowledge of the parameter c. For a cable hanging symmetrically between two fixed points at the same level, the value of c can be determined directly from (5) or (6), provided the central dip of the cable, or

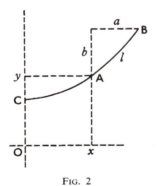

FIG. 2

its total length, are known. But the calculation of c is more trouble-some in the general case illustrated in Fig. 2. Here we are concerned with a cable of given length l hanging between A and B. Let the co-ordinates of A be x and y, and those of B be $(x + a)$ and $(y + b)$, all measured from an origin O below the hypothetical lowest point C as in the previous analysis. Then at A,

$$y = c \cosh \frac{x}{c}, \quad s = c \sinh \frac{x}{c},$$

and at B

$$y + b = c \cosh \frac{x + a}{c}, \quad s + l = c \sinh \frac{x + a}{c},$$

whence, by appropriate subtraction,

$$b = c \left(\cosh \frac{x + a}{c} - \cosh \frac{x}{c} \right),$$

$$l = c \left(\sinh \frac{x + a}{c} - \sinh \frac{x}{c} \right).$$

By squaring these two equations and taking their difference, we have

$$2c \sinh \frac{a}{2c} = \pm (l^2 - b^2)^{\frac{1}{2}}. \tag{8}$$

This gives c in terms of the known dimensions a, b, and l, but it cannot be solved explicitly; numerical solution in any particular case is straightforward, using a table of hyperbolic functions.

2.3. The Parabolic Cable

In many practical suspension bridges the total dead weight of the bridge, instead of being distributed as though uniform along the cables, is more nearly distributed uniformly across the span. Of more practical importance than the common catenary, therefore, is the case of a cable suspended as in Fig. 3 between two points A

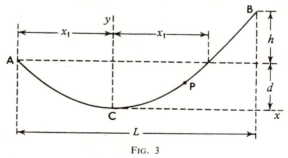

Fig. 3

and B and so loaded (or with a weight per unit length such) that the load per unit of span ($= L$ in the diagram) is constant. Somewhat remarkably, though the catenary was understood at the end of the seventeenth century, this related and simpler problem was not solved till a hundred years later. In 1794 it was proposed to erect a suspension bridge across the Neva, near Leningrad, and it was as a result of considering this proposal that Nicholas Fuss published a solution that year.

We will consider the cable, as before, to be perfectly flexible and inextensible. Let C, the lowest point of the cable, be taken as origin. We will take H as the tension in the cable at C and T as that at any point P. The vertical load on the length of cable CP will be wx (in place of $w's$ for the common catenary), and for equilibrium of this portion of cable

$$T \cos \psi = H, \tag{9}$$

$$T \sin \psi = wx, \tag{10}$$

where ψ is the slope of the cable at P. Dividing (10) by (9), we have

$$\tan \psi = \frac{dy}{dx} = \frac{wx}{H},$$

whence by integration

$$y = \frac{1}{2} \frac{w}{H} x^2 + A,$$

where A is a constant of integration. But at $x = 0$, $y = 0$ by our choice of origin, so that $A = 0$. Thus

$$y = \frac{1}{2} \frac{w}{H} x^2 \tag{11}$$

is the equation of the cable form, which is a parabola with its vertex at C and of latus rectum $2H/w$.

From (9), we have for the tension anywhere

$$T = H \frac{ds}{dx} = H \left\{ 1 + \left(\frac{dy}{dx}\right)^2 \right\}^{\frac{1}{2}},$$

where

$$\frac{dy}{dx} = \frac{w}{H} x.$$

Hence

$$T = H \left\{ 1 + \frac{w^2}{H^2} x^2 \right\}^{\frac{1}{2}}. \tag{12}$$

It remains to determine H in terms of the geometry of the parabola. In the general case illustrated, from (11) and the conditions that $y = d$ at $x = x_1$, and $y = d + h$ at $x = L - x_1$, we find

$$x_1 = \frac{L}{h} \left\{ \sqrt{d(d + h)} - d \right\}, \tag{13}$$

and

$$H = \frac{wx_1^2}{2d}. \tag{14}$$

Hence (12) can be written in the form

$$T = H\left\{1 + \frac{4d^2x^2}{x_1^4}\right\}^{\frac{1}{2}}. \tag{15}$$

In the particular case when the fixed points A and B are at the same level, $x_1 = L/2$ and (14) and (15) become

$$H = \frac{wL^2}{8d} \tag{16}$$

and

$$T = H\left\{1 + \frac{64d^2x^2}{L^4}\right\}^{\frac{1}{2}}. \tag{17}$$

The maximum value for T is then at $x = L/2$ and is

$$T_{\max} = H\left\{1 + \frac{16d^2}{L^2}\right\}^{\frac{1}{2}}. \tag{18}$$

The length of the parabolic cable is in general given by

$$s = \int \left\{1 + \left(\frac{dy}{dx}\right)^2\right\}^{\frac{1}{2}} dx,$$

and in this particular case the total length is therefore

$$l = 2\int_0^{L/2} \left\{1 + \frac{64x^2d^2}{L^4}\right\}^{\frac{1}{2}} dx,$$

$$= \frac{L}{2}\left(1 + \frac{16d^2}{L^2}\right)^{\frac{1}{2}} + \frac{L^2}{8d}\log_e\left\{\frac{4d}{L} + \left(1 + \frac{16d^2}{L^2}\right)^{\frac{1}{2}}\right\}. \tag{19}$$

Alternatively, by expanding the surd in series before integrating, we have the more convenient result

$$l = L\left\{1 + \frac{8}{3}\left(\frac{d}{L}\right)^2 - \frac{32}{5}\left(\frac{d}{L}\right)^4 + \ldots\right\}. \tag{20}$$

For small d/L ratios, it is sufficient to adopt

$$l = L\left\{1 + \frac{8}{3}\left(\frac{d}{L}\right)^2\right\} \tag{21}$$

as adequate for most practical purposes.

Similarly, in the more general case when the ends A and B are not on the same level, if the dip d is measured to the chord line AB, as in Fig. 4, then

$$l = L \left\{ 1 + \frac{8}{3} \left(\frac{d}{L} \right)^2 + \tfrac{1}{2} \tan^2 \alpha \right\}, \tag{22}$$

provided $\tan^2 \alpha$ is small compared with unity.

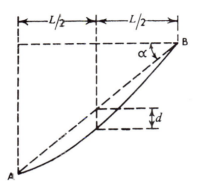

Fig. 4

2.4. The Heterogeneous Cable

In the case of the common catenary, w' was constant measured along the cable; in the case of the parabolic cable, w was constant measured along the span (horizontal) of the cable. It is natural to inquire into the more general case in which w' is a variable, whether measured along the cable or the span. Shortly after solving the common catenary problem, James Bernouilli proceeded to solve this more general problem, inquiring into the law of variation of w' associated with various geometrical forms of cable. We will here note only a few results from this kind of approach.

Following the notation of the previous analysis, and taking our origin in a vertical line through C, the lowest point of the cable, we have as the equations of equilibrium

$$T \cos \psi = H, \tag{23}$$

$$T \sin \psi = \int_0^s w' \, ds. \tag{24}$$

Hence

$$H \tan \psi = \int_0^s w' \, ds.$$

By differentiating this we have

$$w' \cos^2 \psi \, \frac{\mathrm{d}s}{\mathrm{d}\psi} = H \tag{25}$$

as an equation determining w' for a given form of cable.

Applying this result to the two special cases we have already dealt with, we have for the catenary, from (3)

$$\frac{\mathrm{d}s}{\mathrm{d}\psi} = c \sec^2 \psi,$$

whence from (25) above

$$w' = \frac{H}{c},$$

where c is the parameter (constant) of the catenary. Similarly, for a parabola the slope is a constant k times x, so that

$$\tan \psi = kx,$$

whence

$$\cos^2 \psi = \frac{1}{k} \frac{\mathrm{d}\psi}{\mathrm{d}x}.$$

Inserting this in (25), we have

$$w' \frac{\mathrm{d}s}{\mathrm{d}x} = w \sec \psi = kH.$$

Thus w', measured along the cable, must vary so that $w' \sec \psi$, corresponding to w measured along the span, is a constant.

We thus have for the given laws of variation of w' along the cable

    (a) $w' = $ constant, curve $= $ catenary,
    (b) $w' \sec \psi = $ constant, curve $= $ parabola.

A further result of interest is that for

    (c) $w' \cos^3 \psi = $ constant, curve $= $ cycloid.

This last form is of value in the theory of cable oscillations, as we shall see in Chapter 11.

2.5. The Catenary of Uniform Strength

It is clear that a uniform cable loaded only by its own weight, as in the common catenary, is structurally rather inefficient because the tensile stress in the cable, due to the varying tension on the constant sectional area, is not constant. When suspension bridges first became prominent—in this country when Telford's bridge over the Menai Straits was erected—it was natural to inquire into the conditions that would arise with a single cable with its sectional area everywhere proportional to the tension acting along it. David

Gilbert investigated this case for Telford and published his solution in 1826.[31]

Proceeding as for the heterogeneous cable, the equations of equilibrium are

$$T \cos \psi = H, \tag{26}$$

$$T \sin \psi = \int_0^s w' \, ds, \tag{27}$$

whence

$$\tan \psi = \frac{1}{H} \int_0^s w' \, ds. \tag{28}$$

But in order that the stress p due to T may be constant, the sectional area A must vary with T so that

$$\frac{T}{A} = p.$$

But $A = w'/\rho$, where ρ is the weight per unit volume of the cable material, and so

$$T = \frac{p}{\rho} w' = aw', \tag{29}$$

where a has the dimensions of a length.

By substituting for w' from (29) in (28),

$$\tan \psi = \frac{1}{H} \int_0^s \frac{T}{a} \, dx,$$

whence, from (26)

$$\tan \psi = \frac{1}{a} \int_0^s \sec \psi \, ds.$$

By differentiation

$$\sec^2 \psi \frac{d\psi}{ds} = \frac{1}{a} \sec \psi,$$

whence

$$\frac{d\psi}{dx} = \frac{1}{a} \cos \psi = \frac{1}{a} \frac{dx}{ds}.$$

Hence

$$x = a\psi + A,$$

where A is a constant of integration. Taking the origin at the lowest point of the curve, as before, this constant A is zero, and

$$x = a\psi. \tag{30}$$

Then

$$\frac{dy}{dx} = \tan \psi = \tan \frac{x}{a},$$

and by integration

$$y = a \log_e \sec \frac{x}{a}, \tag{31}$$

there being no constant of integration for the origin selected. This is the equation of the catenary of uniform strength as first discovered by Gilbert. The curve is symmetrical about the lowest point and y tends to ∞ as x tends to $\pm\frac{1}{2}\pi a$, where there are vertical asymptotes. Thus the span cannot possibly exceed πa.

This mathematical result needs some qualification in relation to practice, for such a cable of span πa would have an infinite dip. If the dip is limited to a fraction of the span (one-tenth is an average figure for the cables of suspension bridges), (31) gives at once the corresponding value of L/a. Thus, if the dip is d, spans in terms of a for various values of L/d are as tabulated:

TABLE I

L/d	Span L
0	$L = \pi a$
5	$L = 1\cdot 45a$
10	$L = 0\cdot 75a$
15	$L = 0\cdot 52a$

To determine the parameter a for a given material of cable working at a given stress, equation (29) is available; πa will then give the maximum possible span corresponding to that stress. Alternatively, to determine the parameter for a cable of given span and dip, (31) provides a direct numerical equation.

The length of an arc of the cable measured from the origin is given by

$$s = a \log \tan \tfrac{1}{4} \left(\pi + 2\frac{x}{a} \right), \tag{32}$$

and this equation provides a means of determining a when the span and length of the cable are known. In this case it can be shown that

$$\tanh \frac{l}{4a} = \tan \frac{L}{4a}, \tag{33}$$

where l is the length of the cable and L is the span between the fixed points at its ends, assumed level; and a is given by the numerical solution of (33).

Knowing a and the weight per unit length of the cable at any point, the tension there is given at once by (29).

It is of interest to examine the above mathematical results in terms of practical conditions. At the time Gilbert first produced his analysis, the best available material was wrought iron with an ultimate strength of about 25 tons per square inch and a density of about 0·27 lb per cubic inch. From (29), this gives an upper limit for a of 17,000 ft. The mathematical limiting span is π times this, which is 53,000 ft. But this corresponds to the distance between the vertical asymptotes of (31) and assumes that an infinite dip is permissible. If a finite dip of, say, one-tenth of the span is adopted, then from Table I, the practical limit is $0.75 \times 17,000 = 12,700$ ft.

Modern cables are made of steel wire with an ultimate strength of about 100 tons per square inch and a density of 0·28 lb per cubic inch. This corresponds to a value of about 65,000 ft for a and a limiting span on the above basis of $0.75a = 48,700$ ft. But this great span implies a dip of 4870 ft; it would be impracticable with existing materials to build towers of this height, and the cable would have to be regarded as slung between mountains. An alternative approach to the limiting span problem thus arises: assuming a maximum practicable tower height (say 1000 ft) and a value for a of 65,000 ft, what is then the span? Equation (31) gives at once $L = 22,000$ ft. This corresponds to a span to dip ratio of 22.

There is little doubt from the foregoing that the real limits to the spans of suspension cables will be set by other considerations than the academic πa of Gilbert's analysis.

2.6. Comparison of Cable Shapes

The cables of suspension bridges are commonly constructed with a uniform cross-sectional area and thus if allowed to hang freely would adopt the form of the common catenary given by equation (5). But in practice they are often built up *in situ* on a temporary platform, and the roadway is so slung from them by vertical suspension rods that when all is complete and the structure is bearing its own weight, the form of the cables is more nearly parabolic. The aim of this erection procedure is to ensure that so far as the dead weight of the whole bridge (roughly uniform measured along the span) is concerned, it shall be carried wholly by the cables and suspension rods without causing bending actions in any longitudinal girders in the road structure.*

Thus practical interest naturally settles upon the parabolic rather

* Some calculation books of I. K. Brunel, presented by his granddaughter, Lady Noble, to the Library of the University of Bristol, show that in the case of the Clifton Bridge the geometry of the chain cables was based on the common catenary, but that supplementary calculations using the then very recently discovered catenary of uniform strength were also made.

than the catenary form of cable, but there is another reason for this. The shapes of the two curves, for the ratios of span to dip common in suspensions bridges, are so closely alike that to show any differences it is necessary to plot the differences themselves to an exaggerated scale. This has been done in Fig. 5, which indicates

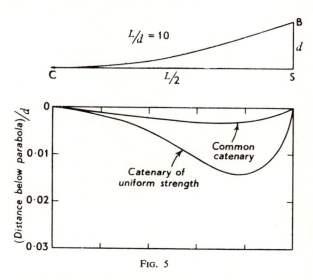

FIG. 5

also that the catenary of uniform strength has much the same shape. And since the cable shapes are alike, the loads in the cable and in any subsidiary structure of a real bridge will also be alike. In these circumstances it is natural to adopt the parabolic shape, with its greater simplicity and familiarity, as the standard one for suspension bridges, and this has become the general custom. In what follows, therefore, in this and succeeding chapters of this book, a parabolic cable will be assumed unless specifically stated otherwise.

2.7. Extensions and Deformations of the Cable

In all the foregoing investigations the cable has been treated as inextensible. To allow for the extension of a cable under load is in effect to allow for the variation of w', the weight per unit length of the cable, with the tension T upon it. This is the problem of the Elastic Catenary first solved by Routh in 1891.[32] Unfortunately the formal solution gives the co-ordinates x and y of the curve adopted by the cable (assumed initially uniform) in terms of the distance s from the origin, measured along the curve in its unextended configuration. The result is thus difficult to apply in any direct manner.

Some more recent work by T. J. Poskitt[33] has recast the basic equations for the elastic catenary in terms of three "elastic catenary functions," and then provided for the rapid numerical solution of the equations by reference to a series of graphs for these functions.

It is easy to show that Routh's solution, which breaks down to the common catenary when Young's modulus for the cable is infinite, also breaks down to the parabola when Young's modulus is very small.* The adoption of the parabola as the practical standard is thus still relevant, particularly if allowance can be made for the overall change of length of the cable under the tensions involved. This procedure, which was probably first introduced by Rankine,[35] will be the one adopted here.

Changes of length of the cable due to the tensions acting on it introduce a change of the cable dip if the ends of the cable are fixed in space. Mathematically, an identical situation can arise from a change of length of cable due to temperature change, or even from the slipping of the cable at an end support. In all such cases we are concerned with the effect of a small change of cable length (measured along the cable) Δl on the dip d. If we can calculate the change in dip Δd, then we can construct the new parabola adopted by the cable.

Returning to para. 2.3, we have from (19) that

$$\frac{\partial l}{\partial d} = \frac{L}{2d}\left\{\left(1 + \frac{16d^2}{L^2}\right)^{\frac{1}{2}} - \frac{L}{4d}\log_e\left[\frac{4d}{L} + \left(1 + \frac{16d^2}{L^2}\right)^{\frac{1}{2}}\right]\right\},$$

whence

$$\Delta l = \frac{L}{2d}\left\{\left(1 + \frac{16d^2}{L^2}\right)^{\frac{1}{2}} - \frac{L}{4d}\log_e\left[\frac{4d}{L} + \left(1 + \frac{16d^2}{L^2}\right)^{\frac{1}{2}}\right]\right\} \Delta d. \quad (34)$$

Alternatively, but to a good approximation, we have by differentiating (20),

$$\Delta l = \frac{16}{15} \cdot \frac{d}{L}\left\{5 - 24\frac{d^2}{L^2} + \ldots\right\} \Delta d. \quad (35)$$

These equations relate the change of length Δl to the change of dip Δd, provided the span L remains constant. If Δl is due to elasticity of the cable, then

$$\Delta l = \int_0^l \frac{T\,ds}{AE}.$$

* Noted by A. S. Ramsey in an example at the end of Chapter XII of *Statics*, 2nd edition, 1943.

By substituting for T from (17) of 2.3 and writing ds in terms of dx, and integrating, we have

$$\Delta l = \frac{Hl}{AE}\left(1 + \frac{16}{3}\frac{d^2}{L^2}\right).\tag{36}$$

Inserting (36) in (35) gives the dip change Δd required.

Similarly, if Δl arises from a change of temperature t, and α is the coefficient of linear expansion of the cable, due to this cause

$$\Delta l = \alpha t l.\tag{37}$$

A related problem arises when the span L itself changes, due, for example, to flexure of the towers supporting the cable. In that case we are concerned with the effect of ΔL on the dip d, the cable length l remaining unchanged. By appropriate partial differentiation of (19), we have

$$\Delta L = \left\{\frac{2\left(1 + \frac{16d^2}{L^2}\right)^{\frac{1}{2}}}{\log_e\left[\frac{4d}{L} + \left(1 + \frac{16d^2}{L^2}\right)^{\frac{1}{2}}\right]} - \frac{L}{2d}\right\}\Delta d,\tag{38}$$

and, as a series approximation

$$\Delta L = \frac{16\frac{d}{L}\left(5 - 24\frac{d^2}{L^2}\right)}{\left(15 - 40\frac{d^2}{L^2} + 288\frac{d^4}{L^4}\right)}\cdot\Delta d.\tag{39}$$

These equations give the change of dip Δd due to a given change of span ΔL when the length of the cable is unaltered.

Having determined the change of dip Δd, whatever the cause, by assuming the cable shape to remain parabolic we can write as before

$$H = \frac{wL^2}{8d},$$

whence

$$\frac{\partial H}{\partial d} = -\frac{wL^2}{8d^2} = -\frac{H}{d}.$$

Thus the resulting change in the cable tension H is

$$\Delta H = -\frac{H}{d}\cdot\Delta d.\tag{40}$$

This relation has more general applications, but nevertheless depends for its accuracy in all cases upon the degree to which the parabolic form is maintained.

CHAPTER 3

The Simple Cable Under Applied Loads

3.1. It was early realised—though this has tended to be overlooked in recent decades—that the behaviour of a suspension bridge under applied loads, even when the bridge is moderately stiffened, depends largely upon the natural response of a heavy cable to applied loads. Thus the first published theory of the stiffened suspension bridge, due to Rankine in 1858, proceeded by giving considerable attention to the mode in which a cable, originally in parabolic form, would change its shape under various types of applied loads; and by 1862 several articles* had appeared which discussed in more detail many of the problems of this chapter. In these later articles the fact that a heavy cable resisted deformation by virtue of its own weight was fully appreciated, and this feature, which Rankine noted but initially ignored,† became of major importance in suspension bridge theory, though it is often masked in modern forms of analysis.

The problems treated by these early workers concerned, firstly, the response of a suspension cable to a single concentrated load (usually at the centre of the span); secondly, the effect of a short uniformly distributed load placed centrally in the span; and, thirdly, the effect of a uniformly distributed load extending from one end of the span to a given distance across it. These cases were chosen for consideration because of their practical design significance, and they are still important on this account. In what follows, special attention will be given to these particular forms of applied load.

---

* W. J. M. Rankine refers (in a footnote in the 2nd edition of his *Manual of Applied Mechanics*) to some unsigned articles in *The Civil Engineer and Architects' Journal* for November and December, 1860. These were followed, probably by the same unknown author, by a series of papers on *The Statics of Bridges* in the same journal, Vol. 25, 1862; it is upon one of these papers that the solution for a single concentrated load given in 3.2 of this chapter is based. In the same year an article appeared in the *Philosophical Magazine* (Vol. XXIII, 1862), by J. H. Pratt, on *The Calculation of the Undulation of an Unstiffened Roadway in a Suspension Bridge as a Heavy Train Passes Over It; and Remarks on the effect of a Suspended Iron Girder in Deadening the Undulation.*

† In Section 9 of Chapter 3 of his *Manual of Applied Mechanics*, Rankine observes: "The weight of the chain itself, being always distributed in the same manner, resists alteration of the figure of the bridge. By leaving it out of account, therefore, an error will be made on the safe side as to the stiffness of the bridge, and calculation will be simplified." In later editions of this work, he qualified the above statement by an interesting footnote.

3.2. The Single Concentrated Load

Consider a single cable, supported between the fixed points A and B shown in Fig. 6, and carrying a uniformly distributed loading

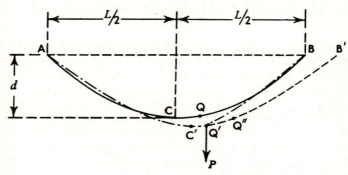

Fig. 6

of w per unit length, measured along the (horizontal) span L. This loading is deemed to include the weight of the cable itself. Let A and B be at the same level, so that the lowest point C will be at the mid-span, at a dip d below A and B.

As indicated in para. 2.3, the cable will hang in parabolic shape with its vertex at C, and if we take this as origin, then

$$y = \frac{1}{2} \cdot \frac{w}{H} \cdot x^2, \tag{1}$$

where H is the tension in the cable at C, and is given by

$$H = \frac{wL^2}{8d}. \tag{2}$$

Suppose now that a vertical force P, small compared with the total uniform load wL, is applied at Q. The cable will deflect in the manner indicated and Q will move to some neighbouring point Q′. Let the position of Q be defined by $x = rL$.

As a result of the application of P, the tension H will change to $(H + h)$ and the equation of the curve AQ′ adopted by the cable AQ will alter from (1) to

$$y_1 = \frac{1}{2} \frac{w}{H + h} x_1^2, \tag{3}$$

where x_1 and y_1 refer to an origin at the vertex C′ of the new parabola, somewhat to the right of the original vertex C. The cable will, of course, hang in a curve that is discontinuous at Q′, so that another expression like (3) will be required to express the geometry of the cable along Q'B, with a new origin to the left of C. The discontinuity at Q' will be such that the vertical components of the cable tension to either side of Q' will together exactly balance the applied force P.

The arc AQ' would, if continued, carry on through the points Q" and B'. Here Q" is chosen so that the horizontal projection of Q'Q" is such that w times this projected length is equal to P. The imaginary arc Q"B' will then clearly be identical in form with the arc Q'B.

The horizontal location of C', the vertex of this parabola AQ'B', may be obtained by noting that, on the actual cable AQ'B, the vertical reactions at A and B are given by

$$V_A = \tfrac{1}{2}wL + P(\tfrac{1}{2} - r), \tag{4}$$

$$V_B = \tfrac{1}{2}wL + P(\tfrac{1}{2} + r). \tag{5}$$

Then since the tension in the cable at C' can have no vertical component,

$$w \times AC' = V_A,$$

or

$$w(\tfrac{1}{2}L + x_0) = \tfrac{1}{2}wL + P(\tfrac{1}{2} - r),$$

where x_0 is the abscissa (measured from C) of the new vertex C'. Hence

$$x_0 = \frac{P}{w}(\tfrac{1}{2} - r). \tag{6}$$

Similarly, for the ordinate y_0 of C' above the origin C, we have for zero moment on the cable at C',

$$(H + h)(d - y_0) = V_A\left(\frac{L}{2} + x_0\right) - \frac{w}{2}\left(\frac{L}{2} + x_0\right)^2,$$

whence

$$y_0 = d - \frac{wL^2}{8(H + h)}\left\{1 + \frac{P}{wL}(1 - 2r)\right\}^2. \tag{7}$$

Thus the coordinates x_0, y_0 of the vertex of the parabola AQ'B' are known, and the whole geometry of the actual cable AQ'B is also known, in terms of the increment h of the horizontal tension H, due to the application of P.

To find h we must discuss the length of the cable. Assuming that any extension of the cable due to the small increment h is negligible, then the essential condition is that the length of the cable in the deformed configuration AQ'B must equal that of the original parabola ACB. If we write v for the vertical deflection of the cable anywhere due to the application of P, and measure this deflection upwards from the original parabolic configuration, then the distribution of v across the span L must be such that the length l of the cable remains constant.

Now

$$l = \int_{-\frac{L}{2}}^{+\frac{L}{2}} \left\{ 1 + \left(\frac{dy}{dx} \right)^2 \right\}^{\frac{1}{2}} dx, \tag{8}$$

and the change Δl due to v may be found by inserting $(y + v)$ for y in (8) and expanding in a Taylor series. If we do this, we have

$$\Delta l = \int_{-\frac{L}{2}}^{+\frac{L}{2}} \frac{(dy/dx)(dv/dx)dx}{\{1 + (dy/dx)^2\}^{\frac{1}{2}}}, \tag{9}$$

neglecting higher terms in v.* If we neglect the departure of the denominator of (9) from unity (justifiable for suspension cables of small dip-to-span ratios), (9) can be written

$$\Delta l = \int_{-\frac{L}{2}}^{+\frac{L}{2}} \frac{dy}{dx} \cdot dv. \tag{10}$$

That this result is approximately true can be seen from the diagram in Fig. 7, where an element MN of the cable is moved to M′N′,

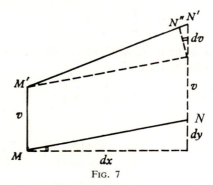

Fig. 7

involving an increase N′N″ in the length of the cable path equal to

$$\frac{dy}{dx} \cdot dv,$$

provided dy/dx is small.

---

* If a further term is included and the denominator of (9) is taken as unity, (9) becomes

$$\Delta l = \int_{-\frac{L}{2}}^{+\frac{L}{2}} \left(\frac{dy}{dx} \right) \left(\frac{dv}{dx} \right) dx + \frac{1}{2} \int_{-\frac{L}{2}}^{+\frac{L}{2}} \left(\frac{dv}{dx} \right)^2 dx.$$

Now in (10), dy/dx refers to the original parabolic shape and, from (1), is

$$\frac{dy}{dx} = \frac{w}{H} \cdot x,$$

whence (10) becomes

$$\Delta l = \frac{w}{H} \cdot \int_{-\frac{L}{2}}^{+\frac{L}{2}} x \, dv. \tag{11}$$

Thus for zero extension

$$\int_0^L x \, dv = 0. \tag{12}$$

Now from A to Q, from the geometry already established, and using C as origin,

$$v = y_0 + \frac{w}{2(H+h)} \left\{ x - \frac{P}{w} (\tfrac{1}{2} - r) \right\}^2 - \frac{w}{2H} x^2, \tag{13}$$

and from Q to B,

$$v = y_0 + \frac{w}{2(H+h)} \left\{ x + \frac{P}{w} (\tfrac{1}{2} + r) \right\}^2$$
$$- \frac{P}{H+h} rL \left(1 + \frac{P}{wL} \right) - \frac{w}{2H} x^2. \tag{14}$$

To use these results in (12), we may write

$$\int_{-\frac{L}{2}}^{+\frac{L}{2}} x \, dv = \int_A^Q x \, dv + \int_Q^B x \, dv = 0,$$

and substitute for dv in terms of dx by differentiating (13) and (14). As a result, we have that condition (12) requires

$$\frac{w^2 L^3}{48 H(H+h)} \left\{ - \frac{4h}{w} + \frac{2H}{w} \frac{P}{wL} (3 - 12r^2) \right\} = 0,$$

whence

$$h = H \frac{3}{2} \frac{P}{wL} (1 - 4r^2). \tag{15}$$

Knowing this increase of tension h, we can find the deflection v anywhere from (13) and (14).

3.3. Some Particular Results for the Single Load

There are a number of interesting results to be drawn from the foregoing analysis for a single load.

(*a*) *Vertical deflections under load*

At the loaded point Q, where $x = rL$, the vertical deflection is given by

$$v_Q = - \frac{P}{wL} \cdot d \cdot \frac{(1 + 12r^2)(1 - 4r^2)}{2 + \dfrac{3P}{wL}(1 - 4r^2)}. \tag{16}$$

The variation of this deflection with r, which expresses the distance of the loaded point Q from C, is plotted in Fig. 8 for values of P/wL

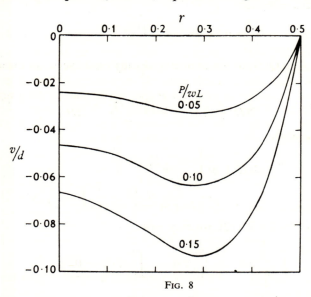

Fig. 8

of 0·05, 0·10 and 0·15. It will be seen from (16) that the relation between v and the load P is not strictly linear. This arises from the presence of P/wL in the denominator, whence it can be seen that as P increases the cable becomes effectively stiffer.* This, of course, agrees with experience, but (16), by the nature of its derivation, is inapplicable to large values of P/wL, corresponding to deflections approaching the order of the dip d. To a good approximation, for small values of P/wL, (16) may be written

$$v_Q = - \frac{P}{2wL} d(1 + 12r^2)(1 - 4r^2), \tag{17}$$

whence by differentiation it is easy to show that a given load will

---

* For a brief but more detailed discussion of the non-linear behaviour of the cable see Ref. 21.

produce a maximum deflection when $r = 1/\sqrt{12}$; that is, when the load is applied at a point distant $0.29L$ from the centre of the span. In this case, we have from (17) that at $r = 1/\sqrt{12}$

$$v_Q = -\frac{2}{3}\frac{P}{wL}\,d. \tag{18}$$

When the load is applied at the centre of the span, $r = 0$ and (17) gives

$$v_0 = -\frac{1}{2}\frac{P}{wL}\,d. \tag{19}$$

(b) Horizontal deflections under load

Equation (10), with appropriate limits, gives the increase, due to the displacements v, in the length along the cable between the end A and any point x in the cable path. If the cable is inextensible, no real increase in its length can occur, and the displacements v result in a horizontal movement u at x towards A. If the depth-to-span ratio of the cable is small, this movement is given closely by (10). Hence, measuring u positive in the direction of x,

$$u = -\int_{x=-\frac{L}{2}}^{x=x} \frac{dy}{dx}\,dv.$$

Then by substituting for dy/dx as before, we have

$$u = -\frac{w}{H}\int_{x=-\frac{L}{2}}^{x=x} x\,dv. \tag{20}$$

Using equation (13) or (14) to perform this integration, we derive for the point Q under the load

$$u_Q = 8 \cdot \frac{P}{w} \cdot \frac{d^2}{L^2} \cdot \frac{r(1-16r^4)}{1+\dfrac{3}{2}\dfrac{P}{wL}(1-4r^2)}, \tag{21}$$

where u_Q is positive measured in the direction of x.

Provided P/wL is small, we may reduce this to

$$u_Q = 8\frac{P}{w}\frac{d^2}{L^2}r(1-16r^4). \tag{22}$$

This will be a maximum when $r^4 = 1/80$; that is, when $r = 0.335L$. In this case, (22) gives

$$u_Q = 2.14\frac{P}{w}\frac{d^2}{L^2}. \tag{23}$$

Were we to explore the horizontal movements of other points in the cable for this loading case, it would be found that all points had moved in the same direction, corresponding to a simple general swinging of the cable in the x-positive direction, that is, towards the applied load. The maximum horizontal movement under any loading will thus tend to occur at the lowest point C of the cable.

When the load is applied at C the centre of the span, $r = 0$ and (22) gives

$$u_C = 0,$$

corresponding to the symmetrical condition then existing.

(c) *Influence coefficients for deflections*

It is sometimes convenient to discuss the deflections of a cable under a given system of loads by means of influence lines, and these can best be recorded by a set of flexibility coefficients representing the deflections of a number of prescribed points along the cable due to unit load at one of them. For many suspension bridge problems it is sufficient to consider loads that are small compared with the weight of the cable and deck (i.e. P/wL small), so that equations such as (17) and (22) apply and response to load may be regarded as linear. In these circumstances the principle of super-position holds and the response of the cable to any system of loading may be determined by adding the effects of separate parts of the system, each of which may be approximately deduced from the influence coefficients.*

As a result of experience in a number of applications, it is found sufficient for many purposes to express the behaviour of the cable by coefficients relating to nine points evenly distributed across its span. We can regard the span as divided into ten equal parts, and take ordinates through the ends of these parts as defining the positions of the nine points in the cable. By a set of calculations based on the foregoing analysis, linearised as indicated, we can construct a table of influence coefficients for deflections, or flexibility co-efficients, as shown in Table II.

As an example, the coefficient of -5.38 given for the deflection at station 4 due to unit load there is derived from (17), thus

$$\frac{v_Q}{d} \text{ (for } r = 0.2) = -\frac{0.1}{2} \times 100(1 + 0.12)(1 - 0.04) = -5.38.$$

It will be noticed that the load positions for maximum vertical displacement occur between stations 2 and 3 and stations 7 and 8,

---

* This procedure was put forward by the author, and its use illustrated by a simple example, in Ref. 18.

where the coefficients approach the value of 6·67 per cent of *d* given by equation (18).

It will be noted that Table II relates to any suspension cable with a dip that is small compared with the span, and (to the order of

TABLE II

FLEXIBILITY COEFFICIENTS FOR VERTICAL DEFLECTIONS DUE
TO VERTICAL LOADS

(unit load $P = 0·1wL$)

Posi-tion of Load	Deflection (percentage of dip) at Station No.								
	1	2	3	4	5	6	7	8	9
1	− 5·28	− 2·95	− 1·07	+ 0·38	+ 1·40	+ 1·98	+ 2·13	+ 1·85	+ 1·15
2	− 2·95	− 6·65	− 3·15	− 0·38	+ 1·60	+ 2·81	+ 3·27	+ 2·94	+ 1·85
3	− 1·07	− 3·14	− 6·22	− 2·30	+ 0·60	+ 2·49	+ 3·38	+ 3·27	+ 2·13
4	+ 0·38	− 0·38	− 2·30	− 5·38	− 1·60	+ 1·02	+ 2·49	+ 2·81	+ 1·98
5	+ 1·40	+ 1·60	+ 0·60	− 1·60	− 5·00	− 1·60	+ 0·60	+ 1·60	+ 1·40
6	+ 1·98	+ 2·81	+ 2·49	+ 1·02	− 1·60	− 5·38	− 2·30	− 0·38	+ 0·38
7	+ 2·13	+ 3·27	+ 3·38	+ 2·49	+ 0·60	− 2·30	− 6·22	− 3·14	− 1·07
8	+ 1·85	+ 2·94	+ 3·27	+ 2·81	+ 1·60	− 0·38	− 3·15	− 6·65	− 2·95
9	+ 1·15	+ 1·85	+ 2·13	+ 1·98	+ 1·40	+ 0·38	− 1·07	− 2·95	− 5·28

accuracy of the analysis used) is expressed in a way that is independent of the ratio of span to dip.* For cables with this ratio in the region of 10 and over, this independence is approximately true and adds to the value of the influence coefficient approach.

The values in the leading diagonal of Table II above may be regarded as expressing the distribution of stiffness of the cable, as measured by a unit vertical load. Thus the cable is stiffer towards the ends (coefficient = − 5·28 at stations 1 and 9) and the centre (coefficient = − 5·00) than near the quarter points of the span (coefficient = − 6·22 at stations 3 and 7). This variation of stiffness, except as the end support points are approached, is not great, and

* It is of interest to compare Table II with that recorded in the Appendix to the Author's paper "Some Experimental Work on Model Suspension Bridges", *Structural Engineer*, Vol. 27, 1949. The latter was based on experimental measurements for a cable with a span-to-dip ratio of 10, made at a time when the approximate independence of the results from this ratio was not realized. The experimental values in the leading diagonal of the table are all lower than the theoretical ones, in the central region by some 6 per cent, and more elsewhere. These differences were large enough to make any effect of the span-to-dip ratio somewhat uncertain.

it is therefore of interest to look at its average value. We can obtain this by the integration of equation (17). This gives for the average stiffness, as a deflection due to a concentrated load P,

$$v_Q = -\frac{8}{15} \cdot \frac{P}{wL} \cdot d, \tag{24}$$

equivalent to a constant value of $-5\cdot33$ along the leading diagonal of Table II.

(d) Energy relationships

It is of interest to examine the behaviour of a cable under a single applied load in terms of energy changes. In an ordinary elastic structure, when a single load P is applied some deflection δ occurs at the load in its line of action, and the applied force does an amount of work $\frac{1}{2}P\delta$; and this work is stored in the structure as elastic strain energy, expressible in terms of the extensions, compressions, and bending actions induced in the structure. But we have assumed our cable to be inextensible, so that the added tension (represented by its horizontal component h) in the cable due to the application of P has not involved any work expressible as strain energy. The same applies to the support points A and B, which have been treated as rigidly fixed. Yet the application of P has involved, to a first approximation, a linearly increasing resistance, so that work has been done by P to the amount $\frac{1}{2}Pv_Q$. Where and how has this been stored?

Now it is clear from expressions such as (17) that the resistance of the cable comes from its initial weight or loading w, and is thus due to gravity. So that the work done by P on the cable must, by the alteration produced in the configuration of the cable, have been stored as potential energy in the gravity field. In other words, the application of P must, on the average, have "lifted" the cable; and for small deflections,

$$\frac{1}{2}Pv_Q = w \int_{-\frac{L}{2}}^{+\frac{L}{2}} v \, \mathrm{d}x. \tag{25}$$

If the results of our approximate analysis are studied in terms of this general relationship, it will be found that the energy balance has not been accurately maintained. This is a common outcome of approximate methods for considering the equilibrium of any structure, and in this particular case can be traced to the approximations involved in the use of (9) or (10) for expressing the inextensibility of the cable.*

* For a detailed examination of the critical questions arising, see Ref. 21.

One further energy matter may be touched upon here. If the cable had in fact been extensible, would this greatly alter the accuracy of the solutions given in this chapter? Let us consider as an extreme case a cable loaded only by its own weight, and with a span-to-dip ratio of 10. Then, to a first approximation, for a cable of density ρ and cross-sectional area A,

$$w = \rho A.$$

If now a central load $P = 0.1wL$ is applied, we have from (19)

$$z_0 = -0.05d = -0.005L.$$

Hence the work done by P is numerically

$$W = \tfrac{1}{2}Pz_0 = 250 \times 10^{-6} \times \rho AL^2. \tag{26}$$

Now the increased tension of the cable will be given approximately by (15), taking $r = 0$. Substituting for H from (2) in (15), we have

$$h = \frac{3}{16}\,\rho AL.$$

Neglecting the variation of the actual tension in the cable corresponding to h, and also the difference between l and L,* we have for the increase of strain energy due to P,

$$U = \frac{h^2L}{2AE},$$

$$= 0.0175 \left(\frac{\rho L}{E}\right) \rho AL^2. \tag{27}$$

If we insert values of $\rho = 0.28$ lb per cubic inch, $L = 3000$ ft, and $E = 30 \times 10^6$ lb per square inch appropriate to steel, in this equation for U, we have

$$U = 5.9 \times 10^{-6} \times \rho AL^2. \tag{28}$$

Thus for such practical conditions U is negligible compared with W.

If we repeat this analysis for a like load at the quarter point ($r = \tfrac{1}{4}$), we obtain in place of (26),

$$W = 330 \times 10^{-6} \times \rho AL^2, \tag{29}$$

and in place of (28)

$$U = 3.3 \times 10^{-6} \times \rho AL^2. \tag{30}$$

It is clear that for the conditions we have considered, U is again negligible compared with W. Our neglect of the extensibility of the cable in most of the problems of this chapter is thus justifiable.†

* For a more accurate analysis see **5.3**, p. 54.
† For a further examination of the effect of extensibility, see Chapter 5, p. 57.

3.4. The Short Distributed Load Centrally Placed

As in para. 3.2, we will consider here a uniform cable carrying, in total, a uniformly distributed dead loading of w per unit length measured along the span. This cable will hang in a parabola with span L and dip d as shown in the diagram.

Suppose now a further loading of p per unit length be applied in the central region of the span, as shown in Fig. 9, having a length

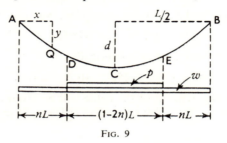

Fig. 9

$(1 - 2n)L$. The cable will now take the form of three parabolic arcs, shown by ADCEB. Here the arcs AD and EB will be similar, but antisymmetrical, corresponding to the loading w, and the arc DE will be different, corresponding to the loading $(p + w)$.

The vertical reactions at A and B, when the superimposed loading p is present, will be equal and given by

$$V_A = V_B = \frac{wL}{2} + \frac{(1 - 2n)pL}{2}. \tag{31}$$

Hence, for a typical point Q in AD, measuring x and y from A/, as origin, we have for zero moment on the cable

$$Hy - V_A x + \tfrac{1}{2}wx^2 = 0, \tag{32}$$

where H is the horizontal reaction at A due to the total system of loading. Substituting from (31) and rearranging, we have for the portion AD

$$y = \frac{wx(L - x) + pLx(1 - 2n)}{2H}. \tag{33}$$

Similarly, for the portion DE

$$y = \frac{(p + w)(L - x)x - pn^2L^2}{2H}. \tag{34}$$

If the central dip, when p is present, is D, then (34) gives

$$H = \frac{L^2}{8D}\left\{(p + w) - 4n^2p\right\}. \tag{35}$$

This, of course, agrees with (16) of para. 2.3 when $p = 0$, for then $D = d$.

We may now conveniently appeal to the approximate equation (22) of para. 2.3 to discuss the lengths of the three parabolic arcs AD, DE and EB, and so, by equating their total length to the original length of the single arc AB, introduce the condition of inextensibility. Now, on this basis, from (33), (34) and (35)

$$l_{AD} = l_{EB} = rL\left\{1 + \frac{3}{3}\cdot\frac{D^2}{L^2}\cdot\frac{n^2w^2}{[(p+w)-4n^2p]^2}\right.$$
$$\left. + 8\cdot\frac{D^2}{L^2}\cdot\frac{[w(1-n)+p(1-2n)]^2}{[(p+w)-4n^2p]^2}\right\}, \quad (36)$$

and

$$l_{DE} = L(1-2n)\left\{1 + \frac{8}{3}\cdot\frac{D^2}{L^2}\cdot\frac{[(p+w)(1-2n)]^2}{[(p+w)-4n^2p]^2}\right\}. \quad (37)$$

But, for zero extension

$$AB = L\left(1 + \frac{8}{3}\frac{d^2}{L^2}\right) = l_{AD} + l_{DE} + l_{EB}. \quad (38)$$

Substituting from (36) and (37) in (38), we have

$$D = d\cdot\frac{\left[\left(\frac{p}{w}+1\right)-4n^2\frac{p}{w}\right]}{\left[\left(\frac{p}{w}+1\right)^2-4n^2\frac{p^2}{w^2}(3-4n)-4n^2\frac{p}{w}(3-2n)\right]^{\frac{1}{2}}}. \quad (39)$$

Having thus found the final dip D in terms of the initial dip d, the deflection v of the cable due to the loading p may be found at any point. In particular, at the centre C, taking v downwards as positive

$$v_0 = D - d. \quad (40)$$

It is a matter of practical interest to determine the length of the loading p that gives rise to a maximum value of v_0. Clearly, as the length $(1 - 2n)L$ grows from a small value, v_0 must also grow, but when $(1 - 2n)L$ approaches L (i.e. when $n \to 0$), v_0 must return to zero, for p then becomes a simple addition to w right across the span. The value of n for a maximum value of v_0 may be found by differentiating the latter with respect to the former, and equating the differential to zero. Using (39) for this purpose, we have for a maximum value of v_0

$$\frac{p}{w}\left(2\frac{p}{w}+1\right)n^3 - 3\frac{p}{w}\left(\frac{p}{w}+1\right)n^2$$
$$+ \frac{3}{4}\left(2\frac{p}{w}+1\right)\left(\frac{p}{w}+1\right)n - \frac{1}{4}\left(\frac{p}{w}+1\right)^2 = 0. \quad (41)$$

Thus n for maximum v_0 depends on the loading ratio (p/w). By inserting a series of values for (p/w) in (41) and solving for n, the results in the following table have been determined.

TABLE III

For Maximum v_0	Loading Ratio p/w				
	$\frac{1}{3}$	$\frac{1}{2}$	1	2	3
n	0·352	0·356	0·375	0·402	0·410
Loaded length $(1 - 2n)$	0·296	0·288	0·250	0·196	0·180
v_0/d	0·022	0·028	0·045	0·066	0·097

These values, it will be noted, are, like the flexibility coefficients in para. 3.3 (Table II), independent of the ratio L/d (to the accuracy of the analysis).

It is of interest to consider the limiting case when the loaded length $(1 - 2n)L$ tends to zero, for this corresponds to the case of a concentrated central load as considered in paras. 3.2 and 3.3. Writing $k = 1 - 2n$, (39) becomes

$$D = d \cdot \frac{1 + 2\left(\frac{p}{w}\right)k - \left(\frac{p}{w}\right)k^2}{\left[1 + 3\left(\frac{p}{w}\right)k + 3\left(\frac{p}{w}\right)^2 k^2 - \left(\frac{p}{w}\right)k^3 + 2\left(\frac{p}{w}\right)^2 k^3\right]^{\frac{1}{2}}}. \quad (42)$$

If we now make $k \to 0$, (42) becomes

$$D = d\, \frac{1 + 2\left(\frac{p}{w}\right)k}{1 + \frac{3}{2}\left(\frac{p}{w}\right)k},$$

whence from (40)

$$v_0 = \frac{1}{2}\left(\frac{p}{w}\right)kd,$$

$$= \frac{1}{2}\frac{P}{wL}d, \quad (43)$$

where P is the total load $(= pkL)$ applied over the very short length $kL[= (1 - 2n)L)]$. This limiting result agrees with (19) in para. 3.3 for the case of a concentrated load P applied at the centre of the span.

Similarly, as $k \to 0$, from (35)

$$H \to \frac{wL^2}{8d}\left[1 + \frac{3}{2}\left(\frac{p}{w}\right)k\right], \quad (44)$$

whence, in the rotation of para. 3.2, the increase of horizontal tension due to P is

$$\dot{n} = H \left(\frac{3}{2} \frac{P}{wL} \right). \qquad (45)$$

This agrees with (15) for a central load, when $r = 0$.

3.5. The Short Distributed Load at One End of the Span

The conditions envisaged in this case are shown in Fig. 10. The applied load, of intensity p per unit length, now stretches from one

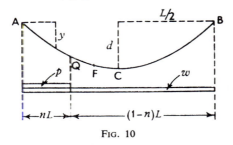

FIG. 10

end A to some point Q only partly across the span AB, and the cable, originally one parabolic arc AB under the loading w, will now deform into two parabolic arcs, one AQ under the loading $(p + w)$, and the other QB under the loading w.

Proceeding as in para. 3.4, the equations for these two arcs are as follows:

For arc AQ,

$$y = \frac{4d}{L^2} \cdot \frac{\left(\frac{p}{w} + 1 \right) x(L - x) - \left(\frac{p}{w} \right) Lx(1 - n)^2}{\left[1 + 2 \left(\frac{p}{w} \right) (3 - 2n)n^2 + \left(\frac{p}{w} \right)^2 n^3(4 - 3n) \right]^{\frac{1}{2}}}. \qquad (46)$$

For arc QB,

$$y = \frac{4d}{L^2} \cdot \frac{\left(\frac{p}{w} n^2 L + x \right) (L - x)}{\left[1 + 2 \left(\frac{p}{w} \right) (3 - 2n)n^2 + \left(\frac{p}{w} \right)^2 n^3(4 - 3n) \right]^{\frac{1}{2}}}. \qquad (47)$$

These two equations correspond to (33) and (34) of para. 3.4, but with H already determined by the condition of inextensibility. In this case, on this condition,

$$H = \frac{wL^2}{8d} \left[1 + 2 \left(\frac{p}{w} \right) (3 - 2n)n^2 + \left(\frac{p}{w} \right)^2 n^3(4 - 3n) \right]^{\frac{1}{2}}. \qquad (48)$$

We are not now so interested in the change of dip at C as in the lowest point of the deformed cable. This lowest point will occur not at C but at some other point F to the left of C. Let x_F be the coordinate of this lowest point.

When $x_F > nL$, F occurs in the arc QB, and by differentiating (47) and equating to zero, we have

$$x_F = \frac{L}{2}\left(1 - \frac{p}{w}\,n^2\right). \tag{49}$$

When $x_F < nL$, F occurs in the arc AQ, and from (46),

$$x_F = \frac{L}{2}\,\frac{1 + \dfrac{p}{w}\,n(2 - n)}{1 + \dfrac{p}{w}}. \tag{50}$$

A special situation arises when F, the lowest point, coincides with the head of the load Q, that is when $x_F = nL$, whence, from (49) or (50),

$$x_F = x_Q = L\left[\sqrt{\left(\frac{w}{p}\right)^2 + \frac{w}{p}} - \frac{w}{p}\right]. \tag{51}$$

Under these conditions F is at its greatest distance from C, as can be deduced by examining (49) and (50). By evaluating $y_F = y_Q$ from (46) and (47), and deducting the original value of y at the same point in the cable, the value of $v_{F,\,Q}$ there can be calculated. The results in Table IV are based on this type of calculation for the values of p/w previously considered.

TABLE IV

For F at Q	Loading Ratio p/w				
	$\frac{1}{3}$	$\frac{1}{2}$	1	2	3
$n = \dfrac{x_F}{L} = \dfrac{x_Q}{L}$	0·464	0·449	0·414	0·366	0·333
$\dfrac{v_{F,\,Q}}{d}$	0·002	0·005	0·015	0·035	0·056

It is interesting to note that the length nL of the span to be covered for this condition of maximum lateral displacement of the cable does not vary much with the loading ratio (p/w) and is, to the accuracy of the analysis, independent of the ratio L/d.

CHAPTER 4

The Rankine Theory

4.1. The first theory of the suspension bridge proper—that is, a bridge comprising a roadway slung from suspension cables and stiffened in some measure by longitudinal girders at the road level —was published by W. J. M. Rankine in 1858.[34] The unstiffened suspension bridges of early times had proved too flexible, particularly for heavy moving loads, and the problem of stiffening such bridges was coming to the fore. Rankine observes:

> It was formerly supposed that, to make a suspension bridge as stiff as a girder bridge, we should use lattice girders sufficiently strong to bear the load of themselves, and that, such being the case, there would be no use for the suspending chains.* But Mr. P. W. Barlow, having made some experiments upon models, finds that very light girders, in comparison with what were supposed to be necessary, are sufficient to stiffen a suspension bridge. If mathematicians had directed their attention to the subject, they might have anticipated this result. The present is believed to be the first investigation of its theory which has appeared in print.

Rankine's theory, started in this way nearly a century ago, somewhat surprisingly still holds a place in many English text-books on the theory of structures, though his name may not always be connected with it. As we shall see later, certain of its ingredients continue to play a part in the more modern *elastic theory* popular in America as a closer approximation. But the current English version of the original theory is now commonly looked upon by English suspension bridge designers as too erroneous to merit serious attention. In its historical setting this view seems rather unfair; when Rankine first produced his theory, suspension bridge designers were concerned with bridges with spans of a few hundred feet, not a few thousand feet. And with these short spans, they thought in terms of deeper girders and lighter moving loads than are common today. It is true that for most purposes, for such conditions a suspension bridge is very uneconomical; an understanding of the economics of bridges had yet to come, as had also many modern materials that have so affected the economic problem.

---

* An echo of the design problems behind Robert Stephenson's Britannia Bridge over the Menai Straits. At one time the tubes of this bridge were to be supported by suspension chains. The tubes were ultimately judged capable of carrying the whole load themselves as simple beams, and the projected chains were never provided. The bridge was opened in 1850, eight years before the publication of Rankine's theory.

But Rankine's theory, provided its grave limitations are made clear, has other than historical importance. It serves as a very simple introduction to some of the essential problems of the stiffened suspension bridge, and without it, the next major theory, the *elastic theory* already mentioned, can hardly be fully understood or set in its proper place in the development of the subject.

4.2. The Two-pinned Stiffening Girder

Contrary to the expositions of this theory in current English textbooks, we will follow Rankine's original treatment by considering first the more natural and common case of a suspension bridge with a stiffening girder pinned at each end, as indicated in the diagram. Following the custom we have adopted in earlier chapters, we will restrict attention for the present to a single span (Fig. 11) with a

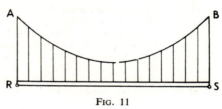

Fig. 11

cable suspended between A and B, and a continuous girder spanning between R and S, where all four points A, B, R and S are regarded as fixed in space. And we will assume that cable and girder are connected by numerous long vertical suspension rods as indicated.

The basic assumptions of the Rankine method are:

(1) That under the total dead loading on the bridge the cable is parabolic and the stiffening girder is unstressed.
(2) That any live loading applied to the girder is so distributed by it to the cable that the latter is called upon to carry a uniformly distributed loading across its whole span.

As a result of (1) and (2), the cable will retain its parabolic shape and so, if inextensible, suffer no deformation, and this is sometimes stated as a basic assumption in place of (2); but this does not seem to have been Rankine's idea* and appears to the writer to lend an air of precision to the theory that was certainly not intended.

With these assumptions the girder RS of our diagram will, when

---

* Rankine appears rather to have regarded (2) as an expression of a design aim that he believed to be approximately achievable. He refers to *a suspension bridge, in which a girder is used to stiffen the bridge. In order that it may do so effectively, any partial or concentrated load on the platform must, by means of the girder, be transmitted to the chain in such a manner as to be uniformly distributed on the chain.*

a given system of live loads is applied, be in equilibrium under the following forces:

(a) The live loads acting vertically downwards.
(b) Vertical reactions V_R and V_S at the pins at R and S, acting either upwards or downwards.
(c) A uniform distributed upward pull from the cable via the suspension rods, acting all along the girder from R to S, of intensity q per unit length.

Of these forces, the magnitudes of V_R, V_S and q are unknown, and only two relevant equations of equilibrium (vertical forces and moments) exist for their determination. Rankine therefore proceeds to make one further assumption:*

(3) That the value of q is equal to the total live load divided by the span L.

With this assumption q can be calculated direct and V_R and V_S found from the two equations of equilibrium. Because of (3), V_R and V_S are necessarily found to be equal and opposite forces.

In this way, the effects of a given system of live loads upon the tension in the cable, or the suspension rod loads, or the bending actions on the stiffening girder, are readily calculable. But it is clear that the method proceeds, as in many engineering approximate methods of analysis, by ensuring equilibrium without making any attempt to check that the displacements involved are compatible. Rankine was, of course, aware of this and did in fact go so far as to check that assumption (3) led to girder deflection modes that were qualitatively such as could reasonably be accommodated by a relatively inextensible cable.

4.3. The Two-pinned Girder with a Single Concentrated Load

Consider the effects of applying at Q in Fig. 12 a single load P to the stiffening girder at a distance x from the end R. Then by assumption (3) above, the uniform loading q between girder and cable is given by

$$q = \frac{P}{L}. \tag{1}$$

---

* This is the assumption mentioned in those few modern text-books that refer to this theory in relation to the two-pinned stiffening girder case. In fact, whilst it was adopted by Rankine for his analysis of the conditions arising under a distributed live load of any length, he was sufficiently aware of its approximate nature as to depart from it when considering a concentrated load at the centre of the span. In that case he argues on physical grounds for a value of q equal to 1·57 times that given by assumption (3).

From the conditions of equilibrium for the girder, with this value of q, we then have

$$V_R = \frac{P}{L}\left(\frac{L}{2} - x\right) = -V_S. \tag{2}$$

The bending moment diagram for the girder is thus given, as in the diagram shown, by the difference between the parabolic curve, with

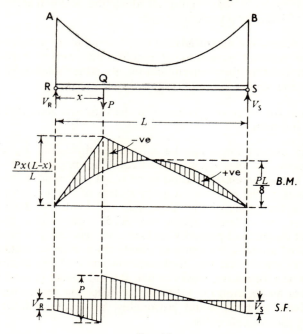

FIG. 12

a maximum ordinate of $qL^2/8 = PL/8$, due to q, and the triangular diagram, with a maximum ordinate of $Px(L - x)/L$. The peak value of the bending moment under the load is

$$M_Q = -\frac{Px(L - x)}{2L} \tag{3}$$

When Q, the point of application of P, is at the centre of the span, $x = L/2$, and

$$M_Q = -\frac{PL}{8}. \tag{4}$$

The variation of shear force on the girder, corresponding to the bending moment diagram, is also shown in Fig. 12.

The foregoing actions on the stiffening girder are accompanied by an increase of tension in the cable due to the uniform pull downward of q per unit span. As in para. 2.3, this uniform loading produces an increased tension measured by its horizontal component h of amount

$$h = \frac{qL^2}{8d} = \frac{PL}{8d},$$ (5)

where d is the dip of the cable. This increase h is, by assumption (3) of para. 4.2, independent of the location of Q along the span.

It is of interest to compare the results (4) and (5) with the results that would obtain were either the cable or the girder absent. Thus, if the girder were called upon to carry the load P alone, as a simple beam

$$M_Q = -\frac{PL}{4},$$

so that the presence of the cable appears to have halved this peak bending action. Alternatively, were the girder absent and the load P applied to the cable alone, then from (15) of para. 3.2,

$$h = \frac{3PL}{16d},$$

which is 50 per cent greater than the value given by (5).

On the basis of this theory, the diagrams in Fig. 13 give influence lines for bending moment and shear force in the stiffening girder at a section distant nL from the end R due to unit rolling load. It will be seen that as soon as the load enters on the span, the bending moment at the given section Z becomes

$$M_Z = +\tfrac{1}{2}L(1 - n)n,$$ (6)

and thereafter falls till the load reaches Z, when its value is

$$M_Z = -\tfrac{1}{2}L(1 - n)n.$$ (7)

Further movement of the load reduces this negative moment till the value reaches that given by (6) when the load is about to leave the span at S. Thus the greatest bending moment arising during the passage of the load is that given by (6) and (7), and when Z is at the centre of the span C the greatest moment is

$$M_C = -\tfrac{1}{8}L,$$ (8)

which agrees, of course, with (4) above.

It is clear from the influence line for shear force that, at any section Z, independent of the value of n, the maximum shear force is $\pm\tfrac{1}{2}$ and arises when the unit load is at the section itself.

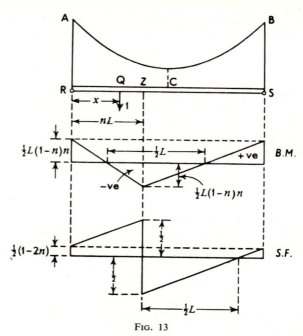

Fig. 13

4.4. The Two-pinned Girder with a Uniform Load

The influence lines just discussed enable the effects of a uniform load, of various lengths and in various positions, to be determined at once. Thus it is evident that, as regards the bending moment at Z, the maximum effect on the girder will occur when a load of length $\tfrac{1}{2}L$ is disposed over the point Z above the *negative* triangle shown. Under these conditions, the moment is given by the area of this triangle, so that

$$M_Z = -\tfrac{1}{8}pL^2(1 - n)n, \tag{9}$$

where p is the loading intensity. This moment M_Z has a maximum value, when Z is at C and $n = \tfrac{1}{2}L$, of

$$M_C = -\tfrac{1}{32}\,pL^2. \tag{10}$$

Had not the cable been present, the greatest bending moment on the beam would have arisen when the whole span was covered, in which case

$$M_C = -\tfrac{1}{8}pL^2. \tag{11}$$

Thus the presence of the cable, on this analysis, reduces the peak value of the bending moment occurring on the girder to one-quarter of the value for a simple beam.

Rankine himself does not appear to have studied the case of partial loading represented by (10), but instead investigated the case of a uniform load extending from one end of the girder (such as R) partly across the span. In this case, from our influence diagram, the maximum moment (negative) occurs when the span is loaded from R over Z to the far end of the *negative* triangle, when

$$M_Z = -\frac{p}{2}\frac{L}{2}(1-n)n \cdot \frac{(1-n)L}{2},$$

$$= -\tfrac{1}{8}pL^2(1-n)^2n. \tag{12}$$

This has a maximum value when $n = \tfrac{1}{3}$, corresponding to $\tfrac{2}{3}$ of the span being covered by the load and

$$M_Z = -\tfrac{1}{54}pL^2. \tag{13}$$

This is Rankine's result, from which he concludes that the girder should have a bending strength $\tfrac{4}{27}$ths of that of a simple beam to carry the same intensity of uniform loading, instead of the $\tfrac{1}{4}$ noted above.

The shear forces at Z due to a uniform load can be studied by the influence diagram for shear in Fig. 13. The maximum positive shear force at Z occurs when the span RZ $(= nL)$ is covered and is

$$F_Z = \tfrac{1}{2}pnL(1-n). \tag{14}$$

When $n = \tfrac{1}{2}$, Z is at the centre of the span and (14) gives

$$F_Z = \tfrac{1}{8}pL. \tag{15}$$

The maximum negative shear force at Z occurs when the span of $\tfrac{1}{2}L$ is covered over the negative triangle to the right of Z, when

$$F_Z = -\tfrac{1}{8}pL, \tag{16}$$

and its value is independent of the position of Z across the span. When the whole span is covered, the influence line diagram gives zero shear at Z, whatever the value of n. This, of course, corresponds to a condition in which p is everywhere balanced by the upward pull q, which by our basic assumption (3) of para. 4.2 is

$$q = \frac{pL}{L} = p. \tag{17}$$

The increased tension h in the cable is then a maximum and is given by

$$h = \frac{pL^2}{8d}, \tag{18}$$

which is the same as that which would arise were the girder absent and p applied to the cable alone.

4.5. The Three-pinned Girder with a Single Concentrated Load

Of the three assumptions of para. 4.2 for the case of a two-pinned girder, if (1) and (2) are adopted for the present case then (3) is unnecessary, for the presence of the central hinge provides a condition of zero moment there for the determination of q. Referring to Fig. 14, for the equilibrium of the hinged girders RC and CS, we have the three equations

$$qL + V_R + V_S - P = 0, \tag{19}$$

$$\tfrac{1}{2}V_R L + \tfrac{1}{8}qL^2 - \tfrac{1}{2}P(L - 2x) = 0, \tag{20}$$

$$\tfrac{1}{2}V_S L + \tfrac{1}{8}qL^2 = 0. \tag{21}$$

Here (20) and (21) refer to the moments about C acting on RC and CS respectively. Solving these equations, we have

$$V_R = P\left(1 - 3\frac{x}{L}\right), \tag{22}$$

$$V_S = -\frac{Px}{L}, \tag{23}$$

$$q = \frac{4Px}{L^2}. \tag{24}$$

The bending moment and shear force diagrams shown in Fig. 14 are based on these results, using the form of presentation already adopted for the two-pinned girder. The peak value of the bending moment under the load is

$$M_0 = -Px\left(1 - 3\frac{x}{L} + 2\frac{x^2}{L^2}\right). \tag{25}$$

When the load is at C, $x = \tfrac{1}{2}L$, and this moment necessarily becomes zero. M_0 has a numerical maximum value when $x = 0.211L$, when

$$M_0 = -0.096PL, \tag{26}$$

which is 0.77 of the corresponding maximum given by (4) for the two-hinged girder. For this particular position of the load, (24) gives

$$q = \frac{P}{L}$$

as in (1) for the two-pinned girder.

The variation of shear force on the girder is shown in the diagram, which is drawn for the case of $x < L/3$. It will be seen from (22) that the sign of V_R changes when x passes through the value of

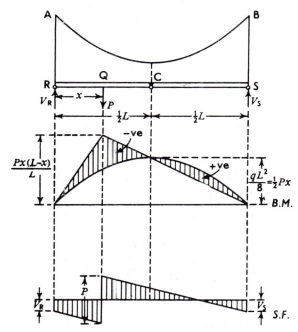

FIG. 14

$L/3$, so that when $x > L/3$ the shear force diagram becomes similar in form to that shown for the two-pinned girder.

In this case of a three-pinned girder, the tension in the cable due to P, instead of being independent of x as in (5), is a function of x because q depends on x as in (24). Thus

$$h = \frac{qL^2}{8d} = \frac{Px}{2d}, \tag{27}$$

and this increase of tension h has a maximum value, when $x = L/2$, of

$$h = \frac{PL}{4d}. \tag{28}$$

As would be expected, the insertion of the central hinge has thrown more of the applied load on to the cable, so that (28) gives a value for h that is twice that for a two-pinned girder.

We can now proceed to construct influence lines for bending and shear actions on the girder, and it will be convenient to restrict attention to values of nL (defining the position of the typical section Z) less than $\frac{1}{2}L$.

Assuming that unit load enters the span at R and proceeds towards Z (Fig. 15), then analysis on the lines of equations (19), (20) and (21) gives for

$$x < nL, \qquad M_Z = -x(1-n)(1-2n), \tag{29}$$

$$x > nL < L/2, \quad M_Z = n\{x(3-2n) - L\}, \tag{30}$$

$$x > L/2, \qquad M_Z = n(L-x)(1-2n). \tag{31}$$

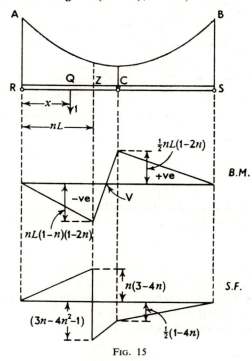

FIG. 15

Thus M_Z varies linearly with x in each case, and the influence line for M_Z is as shown in Fig. 15. Since n is less than $\frac{1}{2}$, the greatest moment at Z will occur when the load is at Z, and is

$$M_Z = -nL(1-n)(1-2n). \tag{32}$$

This is greatest when $n = 0.211$, for which value $M_Z = -0.096L$.

The greatest positive bending moment at Z occurs when the load is at the central hinge, and is given by

$$M_Z = \tfrac{1}{2}nL(1-2n), \tag{33}$$

which has a maximum value of $L/16$ when $n = \frac{1}{4}$. Both these peak bending moments, positive and negative, are considerably less than the value of $L/8$ given by (8) for the two-pinned girder.

The influence line for the shear force at Z is also given in Fig. 15; it is drawn as for values of n between $n = \frac{1}{4}$ and $n = \frac{1}{2}$. For values of n less than $L/4$, the sign of the ordinate at C changes.

4.6. The Three-pinned Girder with a Uniform Load

The effects of a uniform load, of various lengths and in various positions, can now be discussed. As regards the bending moment at Z, it is clear from the appropriate influence line that the maximum negative moment will occur when the span RV (Fig. 15) is covered. From the analysis given in para. 4.5, this length can be shown to be

$$\frac{L}{3 - 2n},$$

whence, from the area of the influence diagram above this base,

$$M_Z = - \frac{pn(1 - n)(1 - 2n)L^2}{2(3 - 2n)}. \tag{34}$$

This has a maximum value for $n = 0.234$, when

$$M_Z = - 0.0188pL^2. \tag{35}$$

The corresponding loaded length is $0.395L$.

Comparison of the result (35) with the similar result (10) for the two-pinned girder, it will be seen that the introduction of the central hinge has reduced somewhat the critical loaded length and has reduced the corresponding greatest negative bending moment by about 40 per cent.

The maximum positive bending moment at Z occurs when the span VS is covered. This has a length of

$$\frac{2(1 - n)L}{3 - 2n},$$

and the corresponding moment at Z is

$$M_Z = + \frac{pn(1 - n)(1 - 2n)L^2}{2(3 - 2n)}. \tag{36}$$

This expression is identical with (34). Hence the greatest positive bending moment occurs at a section distant $0.234L$ from one end and has a value of $+ 0.0188pL^2$; the corresponding loaded length is $0.605L$, measured from the opposite end.

We can now consider the shear forces arising from a uniform load by examining the appropriate influence line. The maximum positive shear force at Z, for the case illustrated (n between $\frac{1}{4}$ and $\frac{1}{2}$), will arise when the span RZ is covered. It can thence be shown that this force has its greatest value when Z is at C ($n = \frac{1}{2}$), when

$$F_C = + \tfrac{1}{8}pL. \tag{37}$$

This is the same as for the two-pinned case. When n is less than $\frac{1}{4}$, however, the greatest positive shear force at Z arises when the spans RZ and XS are covered, where X is a point between Z and C. It can easily be shown that

$$XS = \frac{2L(1 - 2n)}{3 - 4n}, \tag{38}$$

and thence that the greatest positive shear force, in this case, occurs at the end R of the girder ($n = 0$) and is

$$F_R = + \tfrac{1}{6}pL. \tag{39}$$

Dealing with the greatest negative shear force at Z in similar fashion, we find that in the first case (n between $\frac{1}{4}$ and $\frac{1}{2}$) the maximum shear force occurs at $n = \frac{1}{2}$ with a covered length of $L/2$ from S, and has the value

$$F_C = - \tfrac{1}{8}pL. \tag{40}$$

In the second case ($n < \frac{1}{4}$) it occurs at the end R ($n = 0$) with a covered length ZX of $\frac{1}{3}L$ and has a value of

$$F_R = - \tfrac{1}{6}pL. \tag{41}$$

Comparison with the shear force results of para. 4.3 shows that the second case above leads to greater shear forces for the three-pinned girder than for the two-pinned girder.

It is of interest to notice that when the whole span is covered, equations of equilibrium of the type of (19), (20) and (21) give

$$q = p. \tag{42}$$

Thus the loading actions in this case are the same as for the two-pinned girder and the increase in the cable tension is

$$h = \frac{pL^2}{8d} \tag{43}$$

as in (18).

CHAPTER 5

The Elastic Theory

5.1. The development of what is commonly known as the *elastic theory* of suspension bridges grew out of nineteenth century arch theory, and, though the theory later was cast into strain energy form, it still reveals, in terminology and approach, much of its arch origin. Navier, who is regarded on the Continent as the creator of structural analysis as a branch of mechanics,[36] initiated the modern theory of arches in his celebrated *Résumé des Leçons données à l'Ecole des Ponts et Chaussées*, published in 1826, and applied his methods to the design of a suspension bridge over the Seine in Paris, partly constructed in 1831 and later abandoned. Castigliano's later work on arches in terms of his energy theorems[9] led gradually to the restatement of the arch-like theory of suspension bridges in the energy form given by Johnson, Bryan and Turneaure in their comprehensive text-book on *Modern Framed Structures*, first published in 1893. The first English text-book to include a summary of this elastic theory is by A. J. S. Pippard and J. F. Baker.[37]

5.2. In the basic assumptions of this elastic theory, when compared with those of the earlier Rankine theory, is found only one really new feature. Assumptions (1) and (2) of para. 4.2 are made just as before, but in place of (3) relating to the value of q, the uniform distributed loading acting on the cable, is made the assumption that q depends in magnitude (still with uniform distribution) upon the elastic stiffnesses of the cable in tension and the stiffening girder in bending (and to a less extent on the stiffnesses of the towers, etc.). This leads at once to a simple strain energy treatment for the determination of q that is commonly cast in the form of finding the horizontal component of the cable tension. In other words, the cable is treated as an inverted elastic parabolic arch, under uniform loading, with a suspended elastic beam.

This treatment is, of course, essentially one appropriate to a structure that behaves according to Hooke's Law, whereas we have seen in Chapter 3 that the response of a heavy cable to load is in general non-linear. However, for the small deflections commonly concerned in practical suspension bridges, the linear approach is justified and the greater errors arise because q is assumed to be constant across the whole span, irrespective of the distribution of the applied loading. The elastic theory thus represents an advance on Rankine's theory only in the assessment of the magnitude of q, and is essentially an approximate theory that gained popularity in the first half of this century and has proved useful in practice.

5.3. The Two-pinned Girder of a Single Span Bridge with a Single Concentrated Load

To illustrate this method we will consider the simple suspension bridge shown in Fig. 16, and present the analysis in terms of the arch notation that is still customary.

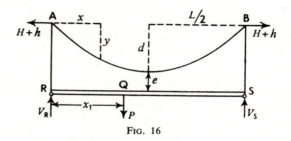

Fig. 16

Let H be the horizontal component of the tension in the cable due to the dead load and h its increase due to the load P at x_1 from R. The bending moment on the beam at any section x will be due partly to P and partly to q, where $q = 8hd/L^2$. For example, at section x

$$M = -V_R x - \tfrac{1}{2}qx^2,$$

$$= -\left\{\frac{P(L - x_1)}{L} - \tfrac{1}{2}qL\right\} x - \tfrac{1}{2}qx^2,$$

$$= -\frac{Px(L - x_1)}{L} + \frac{4hd}{L} x - \frac{4hd}{L^2} x^2,$$

$$= -\frac{Px(L - x_1)}{L} + h\left\{\frac{4d}{L^2} x(L - x)\right\},$$

$$= -\frac{Px(L - x_1)}{L} + hy,$$

where y is the ordinate of the cable at x, as illustrated.

The bending moment M is thus the algebraic sum of that due to the load P and that due to hy. In general, we can write, as is usual for arches,

$$M = \mu + hy, \tag{1}$$

where μ is the moment due to the applied load acting on the girder as if it were isolated and pinned at R and S. The strain energy due to bending of the girder is thus given by

$$U_1 = \int \frac{M^2 dx}{2EI} \tag{2}$$

and for a constant section girder, from (1),

$$\frac{\partial U_1}{dh} = \frac{1}{EI} \int M \frac{\partial M}{\partial h} \, dx$$

$$= \frac{1}{EI} \int (\mu + hy)y \, dx. \tag{3}$$

For our single load case, this equation gives

$$\frac{\partial U_1}{\partial h} = \frac{Px_1 d}{3EI} \left\{ -L + \frac{x_1^2}{L^2} (2L - x_1) \right\} + \frac{8hLd^2}{15EI}. \tag{4}$$

Strain energy will also be stored in the cable due to the change in the tension associated with h. At any point in the cable, this change of tension will be

$$t = h \frac{ds}{dx}, \tag{5}$$

so that this strain energy will amount to

$$U_2 = \int \frac{t^2 ds}{2AE}, \tag{6}$$

where A and E refer to the cable sectional area and modulus of elasticity (assumed the same as for the girder). Hence

$$\frac{\partial U_2}{\partial h} = \frac{2h}{AE} \int \left(\frac{ds}{dx} \right)^2 ds,$$

which for the parabolic cable leads on integration to

$$\frac{\partial U_2}{\partial h} = \frac{hL}{AE} \left\{ \frac{1}{4} \left(\frac{5}{2} + \frac{16d^2}{L^2} \right) \left(1 + \frac{16d^2}{L^2} \right)^{\frac{1}{2}} \right.$$

$$\left. + \frac{3}{32} \cdot \frac{L}{d} \cdot \log_e \left[\frac{4d}{L} + \left(1 + \frac{16d^2}{L^2} \right)^{\frac{1}{2}} \right] \right\}. \tag{7}$$

For simplicity, we will write this as

$$\frac{\partial U_2}{\partial h} = \frac{chL}{AE}, \tag{8}$$

where c refers to the terms in the brackets of (7).

If we wish to allow for the strain energy in the suspension rods, we can treat these as so closely spaced as to be equivalent to a "continuous" set of rods of sectional area a per unit length of the span, where

$$a = \frac{NA_1}{L}. \tag{9}$$

Here N is the actual number of rods, each of area A_1. Thus the stress in a unit spanwise length of the rods will be q/a. The length of a rod at x is, from Fig. 16,

$$e + d\left[1 - \frac{4x}{L^2}(L - x)\right]$$

The strain energy in this element is therefore

$$\delta U_3 = \frac{(q\,dx)^2\left\{e + d\left[1 - \frac{4x}{L^2}(L - x)\right]\right\}}{2a\,dx\,E}, \tag{10}$$

whence, by substituting $8hd/L^2$ for q, integrating with respect to x and differentiating with respect to h,

$$\frac{\partial U_3}{\partial h} = \frac{64d^2h}{L^3aE}\left(e + \frac{d}{3}\right). \tag{11}$$

The effect of the compression in the towers will be small, but can easily be allowed for. The slope of the cable there is

$$\left(\frac{dy}{dx}\right)_0 = \frac{4d}{L},$$

so that the vertical component of the cable tension at the tower (neglecting the effects of backstays, or the like) is

$$V = h\left(\frac{dy}{dx}\right)_0 = \frac{4hd}{L}. \tag{12}$$

Hence, if U_4 is the compression strain energy stored in the two towers, and their cross-sectional areas A_2 are assumed constant over their heights $(d + e)$, then

$$\frac{\partial U_4}{\partial h} = \frac{32d^2(d + e)h}{A_2EL^2}. \tag{13}$$

Adding the results (3), (8), (11) and (13), and equating to zero for minimum strain energy, we have at once an equation for h, thus

$$h = \frac{\dfrac{Px_1d}{3EI}\left\{L - \dfrac{x_1^2}{L^2}(2L - x_1)\right\}}{\dfrac{8Ld^2}{15EI} + \dfrac{cL}{AE} + \dfrac{64d^2}{L^3aE}\left(e + \dfrac{d}{3}\right) + \dfrac{32d^2(e + d)}{A_2EL^2}}. \tag{14}$$

Of the terms in the denominator of (14), the first contributes, for a practical case, about 95 per cent of the total, the second about 5 per cent and the remainder only a fraction each of 1 per cent. Hence to a good approximation we may write

$$h = \frac{\frac{Px_1 d}{3EI}\left\{L - \frac{x_1^2}{L^2}(2L - x_1)\right\}}{\frac{8Ld^2}{15EI} + \frac{cL}{AE}}.$$ (15)

With this value of h, the bending moment anywhere on the girder can be found from (1) and the loading on the suspension rods from

$$q = \frac{8hd}{L^2}.$$

5.4. Examination of Elastic Theory in Terms of the Cable Loading q

It is of interest to examine the results in para. 5.3 in terms of q, the uniform loading on the cable due to P. To do this conveniently, we will assume the term cL/AE in (15) is negligible (i.e. the cable is inextensible) and examine the case when the load P is centrally applied, so that $x_1 = L/2$. Equation (15) then gives

$$h = \frac{25}{128}\frac{PL}{d},$$ (16)

which is independent of the stiffness EI of the girder. This was, of course, a feature of the Rankine theory, for which, under the same conditions, we have already seen ((5) in 4.3)

$$h = \frac{1}{8}\frac{PL}{d}.$$ (17)

Thus (16) is of exactly the same form as (17), but larger in the ratio 25/16. Now q is in both theories related to h thus

$$q = \frac{8hd}{L^2},$$

whence, from (16), the elastic theory gives

$$q = \frac{25}{16}\frac{P}{L}.$$ (18)

This of course accounts for the higher value of h, for the basic assumption of the Rankine theory was

$$q = \frac{P}{L}.$$

It is natural to inquire, with these parallel results before us, just what the strain energy approach represents in physical terms. Now the approximations made in the derivation of (16) correspond to ignoring all the strain energy terms of para. 5.3 except that represented by (4). In other words, (16) corresponds for the case $x_1 = L/2$, to equating (4) to zero; that is, to minimising the bending energy in the girder with respect to h. But q and h are simply proportional to each other, so we see that (16) and (18) both represent the effect of minimising the bending energy in the girder, giving

$$q = \frac{25}{16}\frac{P}{L}.$$

This process, however, could have been done otherwise and more simply. The essential differential $\partial U/\partial h$, and therefore its equivalent $\partial U/\partial q$, represents a deflection; in the latter form a deflection of the girder itself, and we have equated this to zero. But we could have, right at the outset, considered the opposite deflections of the girder due to P and due to q, and realised that the cable, practically inextensible, would ensure that these deflections were approximately equal. For example, for the same case, had we noted that, at the centre,

$$\text{Deflection due to } P = \frac{1}{48}\frac{PL^3}{EI}, \tag{19}$$

$$\text{Deflection due to } q = \frac{5}{384}\frac{qL^4}{EI}, \tag{20}$$

and equated these to give zero deflection there, then we should have found

$$q = \frac{8}{5}\frac{P}{L}. \tag{21}$$

This agrees precisely in form with our strain energy result (18) and differs numerically from it by only 3 per cent. We thus see that the elastic theory, apart from the refinements introduced to make small allowances for the elasticities of cable, towers and suspension rods, is primarily an advance on the Rankine theory in that it provides approximately for the restraining effect of the cable on the girder deflections.

5.5. Influence Lines for Bending Moments and Shear Forces

As in the comparable arch problem, it is convenient to consider influence lines in terms of (1), which can be written

$$M = \left(\frac{\mu}{y} + h\right) y, \tag{22}$$

where, if we concentrate on the section **Z**, the ordinate y is given by

$$y = 4n(1 - n)d. \tag{23}$$

Instead of plotting the influence line for M at **Z** direct, it is customary to plot instead the influence lines for μ/y and h separately, and thence, by algebraic addition, the net influence line for M/y.

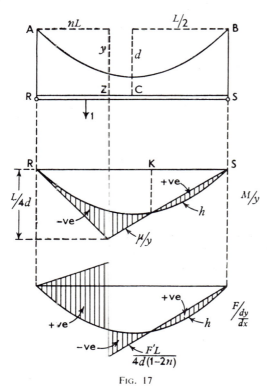

Fig. 17

Proceeding thus, and referring to Fig. 17, the influence line for μ is a triangle with its apex at **Z**, where its ordinate is $n(1 - n)L$; thus the influence line for μ/y is, from (23), also a triangle, the ordinate at **Z** being

$$\frac{n(1 - n)L}{4n(1 - n)d} = \frac{L}{4d}. \tag{24}$$

The influence line for h is the curve obtained from (14) or (15), which it will be seen has the polynomial form

$$h = ax + bx^2 + cx^3, \tag{25}$$

and is symmetrical about the mid-span point C. The difference between this curve for h and the triangle for μ/y gives the influence line for M/y, as shown. It will be seen that a rolling load anywhere between R and K will give a negative bending moment at Z, and between K and S, a positive one. The greatest negative moment at Z arises when the load is at Z; and for the case when Z is at the centre C, this peak moment is

$$M = -\frac{7}{128} L, \tag{26}$$

if $h = \dfrac{25}{128} PL/d$ as in (16). This is only 44 per cent of the result (8) of para. 4.3 for the Rankine theory, a reduction that is roughly typical of bending moment results given by the two theories.

Dealing with shear in the same general manner, we have from (1),

$$F = \frac{dM}{dx} = \frac{d\mu}{dx} + h\frac{dy}{dx};$$

and writing this on the lines of (22), we have

$$F = \left\{\frac{d\mu/dx}{dy/dx} + h\right\}\frac{dy}{dx}. \tag{27}$$

We can thence proceed to discuss the influence line for $F/(dy/dx)$ in terms of the two components $(d\mu/dx)/(dy/dx)$ and h. Writing $d\mu/dx = F'$, the shear corresponding to μ, and noting that, from (23),

$$\frac{dy}{dx} = \frac{4d}{L}(1 - 2n), \tag{28}$$

we have

$$\frac{F}{dy/dx} = \left\{\frac{F'L}{4(1 - 2n)d} + h\right\}. \tag{29}$$

For our single rolling load, this gives the influence line for $F/(dy/dx)$ shown in Fig. 17. Its greatest value occurs when the load is at Z, and for the central section (Z at C)

$$F = F' = \pm \tfrac{1}{2}$$

as for the Rankine case.

5.6. Girder Deflections

We have already seen in para. 5.4 that the elastic theory of this chapter is the first of the suspension bridge theories capable, by its basic concepts, of giving a reasonable estimate of the deflections of the stiffening girder. It is therefore of interest to apply the theory to this problem.

It is convenient to calculate the deflections of the girder directly from the loading actions applied to it. These comprise the external downward loads applied to the girder itself and the uniform upward loading (q) applied by the suspension rods. The deflections due to each of these two systems of loading are readily calculable in the usual way, and their difference gives the net deflection of the girder due to the applied loads.

Three special cases are of general interest: firstly the deflection due to a single concentrated load, secondly that due to a uniformly distributed load over the whole span, and thirdly that due to a uniformly distributed load over one-half of the span.

In the first case, the downward deflection due to a load P at Q distant kL from the end R is given by

$$y' = \frac{PL^3}{6EI}(k-1)\frac{x}{L}\left\{\left(\frac{x}{L}\right)^2 - 2k + k^2\right\} \tag{30}$$

from R to Z, and by

$$y' = \frac{PL^3}{6EI}(k-1)\frac{x}{L}\left\{\left(\frac{x}{L}\right)^2 - 2k + k^2\right\} + \frac{PL^3}{6EI}\left(\frac{x}{L} - k\right)^3 \tag{31}$$

from Z to S. For a central load ($k = \frac{1}{2}$), the central deflection ($x = \frac{1}{2}L$) from (30) or (31) is, of course,

$$y_c' = \frac{1}{48}\frac{PL^3}{EI}. \tag{32}$$

The upward deflection due to q in this case will be given by

$$y'' = \frac{qL^4}{24EI}\frac{x}{L}\left\{1 - 2\left(\frac{x}{L}\right)^2 + \left(\frac{x}{L}\right)^3\right\}, \tag{33}$$

where q, from (15) and neglecting cL/AE, is

$$q = 5\frac{P}{L}k\{1 - 2k^2 + k^3\}. \tag{34}$$

For the special case of a central load ($k = \frac{1}{2}$), these equations give

$$q = \frac{25}{16}\frac{P}{L}$$

as in (18), and at $x = \frac{1}{2}L$

$$y_c'' = \frac{125}{6144}\frac{PL^3}{EI}. \tag{35}$$

Thus, from (32), for the special case of central loading, the net central deflection is

$$y_2 = y_c' - y_c'' = 0.000488\frac{PL^3}{EI}. \tag{36}$$

Similarly, for loading at the quarter points (as at $k = \frac{1}{4}$),

$$y_Q' = 0.01171 \frac{PL^3}{EI},$$

$$q = \frac{285}{256} \frac{P}{L},$$

$$y_Q'' = 0.01033 \frac{PL^3}{EI},$$

whence

$$y_Q = 0.00138 \frac{PL^3}{EI}. \tag{37}$$

It will be seen that the bridge is effectively about three times as flexible at the quarter points as at the centre of the span—an over-estimate but qualitatively a common property of suspension bridges.

In the second case, that of a uniform load of linear intensity p over the whole span, the value of q will, on the approximate assumptions used for deriving (34), be $q = p$; hence no deflections of the girder will occur anywhere along its length. This result, of course, depends primarily on the assumption that the cable is inextensible.

In the third case, if only half the span is covered by a uniformly distributed loading p, the deflection occurs mainly in the region of the quarter points, one rising as much as the other goes down. This quarter point deflection, on the same assumptions as above, is

$$y_Q = 0.000407 \frac{pL^4}{EI}. \tag{38}$$

5.7. Temperature Effects

[This elastic theory, by virtue of its suitability for the approximate discussion of deflections, is also applicable to the problem of temperature effects on a suspension bridge.] In the ideal case under discussion, we have assumed that the cable is fixed in space at its ends A and B, so that any increase in temperature would result, were the cable an isolated one, in an increase of the cable dip to accommodate the increase in its length. The stiffening girder, on the other hand, has in our case been regarded as simply supported at its ends R and S, and so free to expand or contract to meet the effects of temperature changes. Thus, if we consider a cable and girder carrying dead load only, the effect of an increase of temperature will be to reduce the tensions in the suspension rods and so introduce bending moments in the girder corresponding to sagging displacements between R and S.

Now the extension of the cable due to an increase of temperature t will be

$$\Delta l = \alpha t l, \tag{39}$$

where α is the coefficient of thermal expansion for the cable and l is its length, given approximately by (21), in Chapter 2. The corresponding increase of dip at the centre of the span will, from (35) in Chapter 2, be about

$$\Delta d = \frac{3}{16} \alpha t \frac{l^2}{d}. \tag{40}$$

If now it be assumed that the suspension rods do not extend appreciably, due either to changes of load or temperature, then this deflection Δd will be accompanied by a displacement of the girder by the same amount. On the assumption that the uniform loading carried by the suspension rods remains uniform, then the change in this loading q' must be such that

$$\Delta d = \frac{5}{384} \frac{q'L^4}{EI}. \tag{41}$$

Hence

$$q' = \frac{384}{5} \cdot \frac{EI}{L^4} \cdot \Delta d, \tag{42}$$

and the change in H is

$$h' = -\frac{48}{5} \cdot \frac{EI}{dL^2} \cdot \Delta d. \tag{43}$$

From (40), these results may be written

$$q' = \frac{72}{5} \alpha t \frac{EIl^2}{dL^4}, \tag{44}$$

and

$$h' = -\frac{9}{5} \alpha t \frac{EIl^2}{d^2L^2}. \tag{45}$$

The corresponding bending moments in the girder are

$$M = h'y, \tag{46}$$

and at the centre the bending moment is

$$M_C = -\frac{9}{5} \alpha t \frac{EIl^2}{dL^2}. \tag{47}$$

To a first approximation, therefore, since l is not much greater than L,

$$M_C = -\frac{9}{5} \alpha t \frac{EI}{d}. \tag{48}$$

CHAPTER 6

The Application of The Elastic Theory

6.1. In order to make clear the basic assumptions of the Rankine and Elastic Theories, we have in the two preceding chapters considered them only in relation to a single span bridge without side spans of any sort and without any of the many detailed features that in practice commonly complicate the application of any theory. It is not intended in this book to enter into all the questions likely so to arise, but it is clearly desirable to show here how the effects of side spans, which are so common a feature of suspension bridges, can be brought into the framework of our theories. For this purpose the Elastic Theory is the first that can be regarded as at all adequate, and we will restrict attention to it here.

The commonest form of side span is that in which the main cables are continued from the tower tops over the side spans, which are suspended from the cables in the same way as the main span. In such cases the cables are often anchored at the outer ends of the side spans at deck level and are effectively unrestrained horizontally at the tops of the towers, either by the provision of rolling saddles in the case of short wide towers or by the flexibility of the towers themselves in the case of tall slender towers. The side spans, like the main span, are usually freely hinged at their ends to the supports. We will assume all these conditions to obtain and, in addition, concentrate our attention on a geometrically symmetrical case. In other words, we will study the application of the Elastic Theory to the case of a symmetrical three-span suspension bridge with its stiffening girders hinged at the supports, the general form being as shown in Fig. 18.

6.2. The development of the Elastic Theory to modern long-span suspension bridges and the detailed technique of its use owes a great deal to D. B. Steinman[11] and, like other writers and users of the method,* we shall here largely follow his treatment of our sample case and, where possible without violence to our habits elsewhere in this book, his notation.

We assume at the outset that the form specified in Fig. 18 is the initial form of the bridge subject to its own dead weight prior to

---

* For example, Chapter 7 of *An Elementary Treatise on Statically Indeterminate Stresses*, Parcel and Maney, Wiley, 2nd edition, 1936.

the application of any live load. In this initial condition the cable (for convenience we will refer to *one* cable in relation to *one* set of three girders) will be in tension corresponding to a horizontal component H, which, owing to the horizontal freedom at the top of each tower, will be constant over all three spans R_1R, RS and

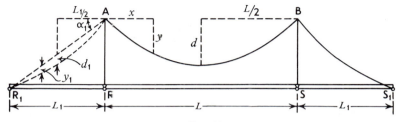

FIG. 18

SS_1. The stiffening girders will be taken to have been so erected that no bending actions exist in them at this stage.

If now any system of live loads is applied to the girders, in one or more of the spans, it will be resisted partly by changes in the tensions in the suspension rods and so by an increase h, which must be uniform, in the horizontal component of the cable tension. Our first problem, therefore, as in the single span case, is to find the value of h for the given system of applied loads.

6.3. Determination of h

We can find h by the strain energy method already outlined in Chapter 5 for the case of a single span. The only modifications required are to allow for the bending strain energy in the side girders, the tension energy (if required) in the cables and suspension rods of the side spans, and the compression energy (if required) in the towers themselves. If we neglect, as is usual, the small contributions to the total strain energy of the towers and suspension rods, the equation (15) of para. 5.3 for the case of a single load P, which reads

$$h = \frac{\dfrac{Px_1d}{3EI}\left\{L - \dfrac{x_1^2}{L^2}(2L - x_1)\right\}}{\dfrac{8Ld^2}{15EI} + \dfrac{cL}{AE}}, \tag{1}$$

becomes for our three span bridge

$$h = \frac{\dfrac{Px_1d}{3EI}\left\{L - \dfrac{x_1^2}{L^2}(2L - x_1)\right\}}{\dfrac{8Ld^2}{15EI} + \dfrac{16}{15}\dfrac{L_1d_1^2}{EI_1} + \dfrac{cL}{A_cE_c} + \dfrac{2c_1L_1}{A_cE_c}}. \tag{2}$$

Here P is assumed to act on the centre span at a distance x_1 from R; the cable is taken to have the same cross-sectional area A_c and Young's modulus E_c throughout;* the girders are assumed to have the same E throughout, but the side spans are taken to have a different effective second moment of area I_1 from the main span girder I. The coefficient c_1 is a non-dimensional term referring to the side span cables in the same way as c refers to the centre span (see equations (7) and (8) of para. 5.3). It is easy to show from (7) of para. 5.3, or otherwise, that to a good approximation

$$c = 1 + 8 \left(\frac{d}{L}\right)^2. \tag{3}$$

Similarly, it can be shown that

$$c_1 = \left\{1 + 8 \left(\frac{d_1}{L_1}\right)^2\right\} \sec^3 \alpha_1, \tag{4}$$

where α_1 is the slope of AR_1 (or BS_1) shown in Fig. 18.

Equation (2) refers to a single load P on the main span. For a more general system of loading the denominator of (2) will still apply, but the numerator will be replaced by terms arising from each of the three spans. On each span, the girder bending moment may, as in (1) of para. 5.3, be written in the form

$$M = \mu + hy,$$

and it will be seen from equations (3) and (4) of para. 5.3 that the numerator of (2) above arises from the evaluation of

$$\int \frac{\mu y \, dx}{EI}$$

for the centre (loaded) span. Thus for a general loading system, equation (2) for h will become

$$h = \frac{\dfrac{1}{E}\left\{\displaystyle\int_0^L \frac{\mu y \, dx}{I} + \int_0^{L_1} \frac{\mu_1' y_1 dx}{I_1} + \int_0^{L_1} \frac{\mu_1'' y_1 \, dx}{I_1}\right\}}{\dfrac{8Ld^2}{15EI} + \dfrac{16L_1 d_1^2}{15EI_1} + \dfrac{cL}{A_c E_c} + \dfrac{2c_1 L_1}{A_c E_c}}, \tag{5}$$

where μ_1' and μ_1'' refer to bending moments on the two side spans due to the loads thereon as if the cables were not present and the girders were simply supported beams.

---

* E_c is for steel ropes commonly appreciably less than E for steel itself, due to the twist in the wires; parallel wire cables do not suffer from this disability.

Steinman has systematised the calculation of h from (5) for a number of practically important loading cases, as follows:

(*a*) Single load P on main span at distance kL from one end.

In this case, (2) applies and Steinman writes

$$h = \frac{1}{N\dfrac{d}{L}} . B(k) . P, \tag{6}$$

where N is the denominator of (2) and the function

$$B(k) = k(1 - 2k^2 + k^3). \tag{7}$$

This function is derived directly from the numerator of (2) by writing $k = x_1/L$; its value for a series of values of k has been tabulated by Steinman and is given in Table V below.

(*b*) Single load P on side span at distance $k_1 L_1$ from one end.

In this case, Steinman gives

$$h = \frac{1}{N\dfrac{d}{L}} . \frac{IL_1^2 d_1}{I_1 L^2 d} . B(k_1)P, \tag{8}$$

where $B(k_1)$ is the same function of k_1 as (7) is of k.

(*c*) Uniform loading of p per unit length acting on main span from one end to a distance kL therefrom.

By integrating (6) or otherwise,

$$h = \frac{1}{5N\dfrac{d}{L}} . F(k) . pL, \tag{9}$$

where the function

$$F(k) = \frac{5}{2} k^2 - \frac{5}{2} k^4 + k^5, \tag{10}$$

and is evaluated in Table V, p. 68.

(*d*) Uniform loading of p_1 per unit length acting on side span from one end (as R_1) to a distance $k_1 L_1$ therefrom.

In this case

$$h = \frac{1}{5N\dfrac{d}{L}} . \frac{IL_1^3 d_1}{I_1 L^3 d} . F(k_1)p_1 L. \tag{11}$$

Here $F(k_1)$ is the same function of k_1 as (10) is of k.

(*e*) Uniform loading of p per unit length acting on the main span from $x = iL$ to $x = kL$.

By applying (9) to derive this case, we have

$$h = \frac{1}{5N\dfrac{d}{L}} \cdot [F(k) - F(j)]pL. \tag{12}$$

Under conditions of uniform loading, the maximum value of h

TABLE V

STEINMAN'S FUNCTIONS FOR APPLICATIONS OF ELASTIC THEORY*

k	$B(k)$	$C(k)$	$D(k)$	$F(k)$	$G(k)$
0·00	0·0000	0·0000	2·0000	0·0000	0·4000
0·05	0·0498	0·0524	1·7511	0·0062	0·4404
0·10	0·0981	0·1090	1·5090	0·0248	0·4816
0·15	0·1438	0·1691	1·2790	0·0550	0·5232
0·20	0·1856	0·2320	1·0650	0·0963	0·5648
0·25	0·2227	0·2969	0·8704	0·1474	0·6062
0·30	0·2541	0·3630	0·6962	0·2072	0·6472
0·35	0·2793	0·4296	0·5445	0·2740	0·6874
0·40	0·2976	0·4960	0·4147	0·3462	0·7264
0·45	0·3088	0·5614	0·3065	0·4222	0·7640
0·50	0·3125	0·6250	0·2188	0·5000	0·8000
0·55	0·3088	0·6861	0·1497	0·5778	0·8340
0·60	0·2976	0·7440	0·0973	0·6538	0·8656
0·65	0·2793	0·7979	0·0593	0·7260	0·8946
0·70	0·2541	0·8470	0·0332	0·7928	0·9208
0·75	0·2227	0·8906	0·0166	0·8526	0·9438
0·80	0·1856	0·9280	0·0070	0·9037	0·9632
0·85	0·1438	0·9584	0·0023	0·9450	0·9788
0·90	0·0981	0·9810	0·0005	0·9752	0·9904
0·95	0·0498	0·9951	0·0003	0·9938	0·9976
1·00	0·0000	1·0000	0·0000	1·0000	1·0000

will, of course, arise when all three spans are covered. In this case, from (9) with $k = 1$ and (11) with $k_1 = 1$, we have

$$q_{max} = \frac{1}{5N\frac{d}{L}}\left\{1 + 2\frac{IL_1^3d_1}{I_1L^3d}\right\}pL, \tag{13}$$

where $p = p_1$ is the intensity of the loading throughout. If we consider as typical a case in which $L_1 = \frac{1}{2}L$, $I_1 = I$, $d_1 = \frac{1}{2}d$, the second term in the brackets of (13) takes the value of $1/16$; it is thus evident that the presence of the side spans has little effect upon the design load for the cable.

6.4. Bending Moments in Stiffening Girders

As in para. 5.5, it is convenient to consider the bending moment on any span in the form

$$M = \mu + hy,$$

and to construct influence lines for each span in terms of

$$\frac{M}{y} = \left(\frac{\mu}{y} + h\right).$$

We thus have the influence diagram of Fig. 19 (*a*) for the moments at Z in the main span due to unit rolling load, and the diagram of Fig. 19 (*b*) for the moments at Z_1 in the side span.

Considering Fig. 19 (*a*) first, the position of the critical point K, corresponding to the intersection of the μ/y and h curves in the main span, can be deduced from (24) and (25) of para. 5.5, and is given by

$$C(k) = k + k^2 - k^3 = N\frac{d\,x_1}{L\,y_1} = \frac{N}{4}\frac{L}{L-x_1}. \tag{14}$$

For the ready determination of k from this equation for a given section Z defined by x_1, the value of the function $C(k)$ is tabulated in Table V. The maximum positive moment at Z will occur when the side spans R_1R and SS_1 are loaded together with the portion KS of the main span. For uniform loading p, this peak moment is

$$M = +\frac{2px_1(L-x_1)}{5N}\left\{D(k) + 4\frac{IL_1^3d_1}{I_1L^3d}\right\}, \tag{15}$$

where the function

$$D(k) = (2 - k - 4k^2 + 3k^2)(1 - k)^2 \tag{16}$$

is given in Table V, and the value of k concerned is that deduced from (14).

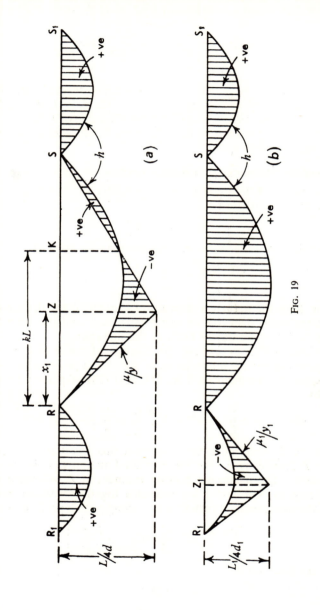

Fig. 19

70

The maximum negative moment at Z will occur when the portion RK of the main span is loaded and all other portions of the spans are unloaded. In this case, the peak moment at Z is

$$M = - \tfrac{1}{2}px_1(L - x_1)\left\{1 - \frac{8}{5N}[1 - \tfrac{1}{2}D(k)]\right\}. \qquad (17)$$

In the case when Z is near the centre of the main span, the μ/y curve of Fig. 19 (a) may cut the h curve in two points instead of one only; in that case the loaded portions of the spans for the peak moments of (15) and (17) will be modified and the formulae altered accordingly. It will be found, however, that the function $D(k)$ is still relevant.

Considering now Fig. 19 (b) for the side spans, no critical points arise, and for a uniform loading p the greatest positive moment at Z_1 occurs when the other two spans RS and SS_1 are loaded; in that case, the peak moment at Z_1 is

$$M_1 = y_1 \left\{ \frac{1 + \dfrac{IL_1{}^3d_1}{I_1L^3d}}{5N\dfrac{d}{L}} \right\} pL. \qquad (18)$$

For the greatest negative moment at Z_1, only the span R_1R must be loaded, when

$$M_1 = - y_1 \frac{L_1{}^2}{8Ld_1}\left\{1 - \frac{8}{5N}\frac{IL_1d_1{}^2}{I_1Ld^2}\right\} pL. \qquad (19)$$

In practice, by applying (15), (17), (18) and (19), it is found, as would be expected, that the greatest peak moments on the main span tend to occur in the $\frac{1}{4}$ span regions, and on the side spans, in the centre span regions. For side spans each about half the main span, these peak moments are all of the same order.

6.5. Shear Forces in Stiffening Girders

Following the treatment of shear in para. 5.5, it is convenient to treat it in terms of $d\mu/dx = F'$ and of h, and so construct influence lines for each span for

$$F\left/ \frac{dy}{dx} \right. = \left\{ \frac{F'L}{4(1 - 2n)d} + h \right\}, \qquad (20)$$

where $n = x_1/L$. On this basis, the influence line for $F/(dy/dx)$ at Z for unit load in the main span is given in Fig. 20 (a) and for unit load in a side span in Fig. 20 (b).

As for the bending moments, the shears (positive and negative) at Z due to loads in various parts of the three spans can be deduced

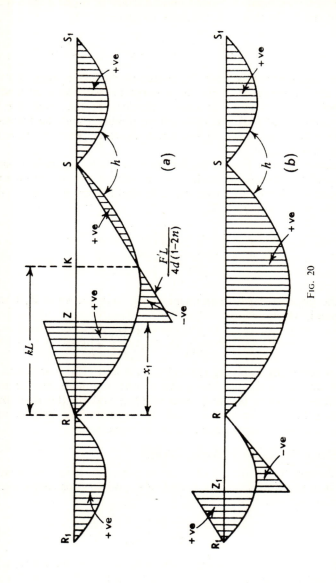

$$\frac{F'L}{4d}(1-2n)$$

FIG. 20

72

from these two diagrams, in which a critical or intersection point K can occur in the first. The position of this point is given by

$$C(k) = k - k^2 - k^3 = \frac{N}{4} \frac{L}{L - 2x_1}, \tag{21}$$

where the function $C(k)$ is tabulated in Table V. In this case, for uniform loading the maximum negative shear at Z occurs when the length ZK of the main span only is covered by the loading, and is given by

$$F = \tfrac{1}{2}pL \left(1 - \frac{x_1}{L}\right)^2 \left\{1 - \frac{8}{N} \left(\tfrac{1}{2} - \frac{x_1}{L}\right) G\left(\frac{x_1}{L}\right)\right\}$$
$$- \tfrac{1}{2}pL(1 - k)^2 \left\{1 - \frac{8}{N} \left(\tfrac{1}{2} - \frac{x_1}{L}\right) G(k)\right\}, \tag{22}$$

where the function $G(x_1/L)$ or $G(k)$ is defined by

$$G(k) = \tfrac{2}{5} (1 - k)^3 - (1 - k)^2 + 1. \tag{23}$$

Values for this function are included in Table V. When no intersection occurs at K, the maximum negative shear at Z occurs when the length ZS is covered, in which case the shear force at Z is given directly by the first of the two terms in (22). The maximum positive shear corresponding to these two conditions occurs when the side spans are covered and either ZK or ZS is covered in the main span. The corresponding value of the shear force at Z is conveniently obtained by deducting the shear given by (22), or by its first term only, from the shear at Z when all spans are fully covered. The value of the latter is

$$F = \tfrac{1}{2}pL \left(1 - 2\frac{x_1}{L}\right) \left\{1 - \frac{8}{5N} \left(1 + \frac{2IL_1{}^3 d_1}{I_1 L^3 d}\right)\right\}. \tag{24}$$

Peak negative and positive side span shears can be obtained similarly. The maximum negative shear at Z_1, which occurs when only Z_1R is covered by the load, is given by

$$F = \tfrac{1}{2}pL_1 \left(1 - \frac{x_1}{L}\right)^2 \left\{1 - \frac{8}{N} \frac{IL_1 d_1{}^2}{I_1 L d^2} \left(\tfrac{1}{2} - \frac{x_1}{L_1}\right) G\left(\frac{x_1}{L_1}\right)\right\}, \tag{25}$$

and the shear at Z_1 when all spans are fully covered is given by

$$F = \tfrac{1}{2}pL_1 \left(1 - 2\frac{x_1}{L_1}\right) \left\{1 - \frac{8}{5N} \frac{d_1 L^2}{dL_1{}^2} \left(1 + 2\frac{IL_1{}^3 d_1}{I_1 L^3 d}\right)\right\}. \tag{26}$$

6.6. Cable Movement at Top of Tower

Whether the cables rest on saddles rolling on the tops of the towers or are fixed there to flexible towers, there will be a tendency, under any general system of loading, for the cable at a tower top to move horizontally. For uniform loading, such movement will clearly be greatest in an outward direction when one side span only is loaded, and in an inward direction when one side span and the main span are loaded. The latter condition will give rise to the peak value of the horizontal movement, and can be evaluated quite simply by considering the upward deflection of the unloaded side span girder under the upward loading q_1 from the cable above it. Let h be the horizontal component of the tension in this cable due to the uniform loading covering the main span and the other side span. Then the corresponding value of q_1 will be

$$q_1 = \frac{8hd_1}{L_1^2},$$

and the deflection at the centre of the unloaded side span—a change of the dip d_1—will be

$$\Delta d_1 = \frac{5}{384} \frac{q_1 L_1^4}{EI_1}$$

$$= \frac{5}{48} \frac{d_1 L_1^2}{EI_1} \cdot h. \tag{27}$$

Assuming the cable is inextensible, this change of dip Δd_1 will be accompanied by a horizontal movement Δ at the top of the tower. To a first approximation, because the inclination α_1 of the parabolic base line AR_1 of the side span is usually small, this movement Δ is given by (39) of Chapter 2, whence for small values of d_1/L_1,

$$\Delta = \frac{16}{3} \frac{d_1}{L_1} \times \frac{5}{48} \frac{d_1 L_1^2}{EI_1} h = \frac{5}{9} \frac{d_1^2 L_1}{EI_1} h. \tag{28}$$

A more accurate result is obtained by multiplying this result by $\cos^2 \alpha_1$.*

In determining this movement Δ we have ignored any resistance due either to friction of the saddle or to bending of the tower. Both are usually inadequate to produce much effect; in the case of a flexible tower fixed at its base and with the cable attached firmly to its top, it is usual to design the tower to withstand safely in bending the whole of the movement Δ impressed at its top.†

---

* The full equation for ΔL_1 in terms of Δd_1 for a side span is given by Steinman on p. 146 of his *Practical Treatise on Suspension Bridges*.[11] Johnson, Bryan and Turneaure,[38] give a constant of $\frac{8}{15}$ in (28) instead of $\frac{5}{9}$; they base it on equation (4) of our Chapter 5, with $p = 0$, but this assumes that the vertex of the parabola of the side span cable coincides with the end R_1.

† See Chapter 13.

CHAPTER 7

The Deflection Theory

7.1. While the elastic theory of suspension bridges was developing on strain energy lines, an essentially more advanced theory was being put forward by Melan[10] in terms of the differential equations of the cable and stiffening girder. As in the case of the elastic theory, the treatment followed the familiar arch form and the concentration of effort was in the first case directed to the determination of the increment h in the horizontal component of the cable tension due to some live loading p per unit length on the girder.

To understand the essentially new feature introduced by Melan, it is convenient to restrict attention to a single span bridge and to consider the bending moment M on any section of the stiffening girder. As we saw in (1) of para. 5.3, this can conveniently be written as

$$M = \mu + hy, \tag{1}$$

where μ is the bending moment, due to the applied load, on the girder when treated as isolated and simply supported at its ends. But in deriving (1), it was implicitly assumed that the deflection v of the girder was negligible compared with the ordinates y of the initial cable shape. If this is not so, allowance must be made for the corresponding increase v of the ordinate y.* To do this, we may note from the derivation in para. 5.3 of (1) that the second term represents the effect of the suspension rod loading. Now, in general, this will have two parts, one w due to dead load† and the other q due to live load. The moment term Hy due to w will be exactly balanced by μ_1, due to the dead loading, so that the initial moment M_1 on the girder is zero. That is, for the dead loading we may write

$$M_1 = \mu_1 + Hy = 0. \tag{2}$$

When the live loading is added, if we neglect v, this expression (2) becomes

$$M = \mu_1 + \mu + (H + h)y,$$
$$= \mu + hy,$$

---

* We here continue to assume that the suspension rods are inextensible, and the vertical deflections of cable and girder are thus everywhere equal.

† It is convenient to assume here and throughout this chapter that the total dead load w is concentrated at the level of the girder.

as in (1). If, however, we treat v as sensible compared with y, then

$$M = \mu_1 + \mu + (H + h)(y + v),$$
$$= \mu + hy + (H + h)v. \tag{3}$$

We see at once that the presence of v gives rise to a change of M that, because of the multiplier $(H + h)$, may be appreciable even when v is small compared with y. It was the realisation of this that made Melan seek to allow for v *ab initio* in his theory and led to the term *Deflection Theory* for the analysis he developed.

7.2. Basic Equations

If we consider the initial shape of the stiffening girder as that of the dead loading condition, and measure the deflections of the girder from this condition, these deflections will be wholly due to the live loading p (which may be variable with x along the span) and the induced suspension rod loading q. Assuming the ordinary theory of bending to apply, we may therefore write for the flexure of the girder

$$EI \cdot \frac{d^4v}{dx^4} = p - q. \tag{4}$$

On the same lines, we may seek to write down the differential equation for the cable form. Now we may note from Fig. 21 that

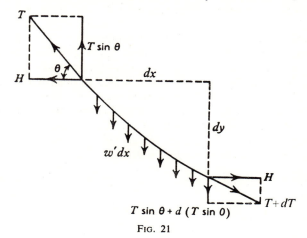

FIG. 21

for the balance of vertical forces on the typical element of cable shown,

$$d(T \sin \theta) + w' dx = 0, \tag{5}$$

where w' is the loading on the cable. But since

$$T \sin \theta = H \tan \theta,$$

and H is constant,

$$\frac{d}{dx}(T \sin \theta) = H \cdot \frac{d}{dx}(\tan \theta),$$

$$= H \cdot \frac{d}{dx}\left(\frac{dy}{dx}\right),$$

$$= H \cdot \frac{d^2 y}{dx^2}.$$

Thus (5) becomes

$$H \cdot \frac{d^2 y}{dx^2} = -w'. \tag{6}$$

But in our case, the loading w' on the cable that is involved is the loading $(w + q)$ provided by the suspension rods, and the ordinates y have, due to q, increased to $(y + v)$. Moreover, H has increased to $(H + h)$. Expression (6), when applied to the cable of our suspension bridge, may therefore be written, for the live load condition,

$$(H + h)\frac{d^2}{dx^2}(y + v) = -w - q, \tag{7}$$

and for the dead load condition

$$H \cdot \frac{d^2 y}{dx^2} = -w. \tag{8}$$

Substituting from (8) in (7) to eliminate w gives

$$h \cdot \frac{d^2 y}{dx^2} + (H + h)\frac{d^2 v}{dx^2} = -q. \tag{9}$$

We now have two equations (4) and (9) relating respectively to the girder and the cable in terms of the unknowns h, v and q. Combining these, by substituting from (9) in (4) to eliminate q, we have

$$EI \cdot \frac{d^4 v}{dx^4} - (H + h)\frac{d^2 v}{dx^2} = p + h\frac{d^2 y}{dx^2}. \tag{10}$$

This equation, in the unknowns h and v, is the fundamental differential equation of the suspension bridge in its classical form.*

* See, for example, Chapter VII, Section 5, *Mathematical Methods in Engineering*, T. von Karman and M. A. Biot. McGraw-Hill, 1940.

The same result can, of course, be expressed more generally in terms of μ. Taking (3) and writing

$$M = EI \cdot \frac{d^2 v}{dx^2},$$

we have

$$\frac{d^2 v}{dx^2} = \frac{\mu + h\,y}{EI} + \frac{H + h}{EI} \cdot v,$$

$$= c^2 v + \frac{c^2}{H + h}\,(\mu + h\,y), \tag{11}$$

where

$$c^2 = \frac{H + h}{EI}. \tag{12}$$

This equation (11) is the form used by Steinman.*

7.3. The basic equation (10) or (11) has two unknowns and before our problem is solved another relation between h and v is required. This is given by the conditions governing the extension of the cable and its overall length.

Now it is clear that, to a first approximation, the elastic extension of the cable due to the applied loading will be

$$\Delta l = \frac{hl}{AE}. \tag{13}$$

The degree of accuracy of this expression can be judged by the more accurate results in (36) of para. 2.7 and (8) of para. 5.3. This extension, which in practice will be very small and often negligible, must match any shortening arising from the deflection v along the cable. This shortening has already been discussed in para. 3.2, and is given to a close approximation by (11) there. Repeating this for our case, we have

$$\Delta l = \frac{w}{H} \int_0^L x \, dv, \tag{14}$$

and, on integrating by parts,† noting that v is zero at both limits

$$\Delta l = -\frac{w}{H} \int_0^L v \, dx. \tag{15}$$

Combining (13) and (15), we have for fixed ends to the cable

$$\Delta l = \frac{hl}{AE} - \frac{w}{H} \int_0^L v \, dx = 0. \tag{16}$$

---

* Apart from a different sign convention for μ. See p. 249 of his *Practical Treatise on Suspension Bridges*, 2nd edition, 1929.

† We are here treating v as a continuous function of x; for the errors involved in this in some cases, see Ref. 21.

In many practical cases, the elastic extension of the cable is negligible and the condition (16) reduces to

$$\int_0^L v \, dx = 0. \tag{17}$$

The problem of the suspension bridge, cast in these terms, thus reduces to the simultaneous solution of (10) or (11) and (16) or (17).

7.4. Case of Uniform Loading

It will be convenient to develop the theory further in terms of a particular case of general interest. Let us assume that p is a uniform loading distributed from $x = 0$ to $x = kL$; then by integrating (11) twice, we have:

$$v = \frac{h}{H + h} \left\{ C_1 e^{cx} + C_2 e^{-cx} + \left(\frac{u}{h} + y \right) + \frac{1}{c^2} \left(\frac{8d}{L^2} - \frac{p}{h} \right) \right\}, \tag{18}$$

where C_1 and C_2 are constants of integration to be determined by the boundary conditions

$$x = 0, \qquad v = 0,$$
$$x = L, \qquad v = 0,$$

and continuity conditions at $x = kL$. Inserting these conditions, we find for the loaded span $x = 0$ to $x = kL$,

$$C_1 = \frac{\dfrac{p}{2hc^2} \left\{ e^{cL(1-k)} + e^{-cL(1-k)} - 2e^{-cL} \right\} - \dfrac{8d}{c^2L^2}(1 - e^{-cL})}{e^{cL} - e^{-cL}}, \tag{19}$$

$$C_2 = -C_1 - \frac{1}{c^2} \left(\frac{8d}{L^2} - \frac{p}{h} \right), \tag{20}$$

and for the unloaded span $x = kL$ to $x = L$,

$$C_1' = C_1 - \frac{p}{2hc^2} e^{-kcL}, \tag{21}$$

$$C_2' = -C_1' e^{2cL} - \frac{8d}{c^2L^2} e^{cL}. \tag{22}$$

These equations give direct expressions for v in terms of known parameters except for h, present explicitly and also implicitly in c. Owing to the fact that h depends on p, it is clear from (18) that v is not strictly proportional to the applied loading p. Thus, in general, the bridge behaves non-linearly, and the principle of superposition, and the method of influence lines based thereon, cannot be applied with precision.

For the evaluation of h itself, we must turn to (16) or (17). By inserting (18) in the equation (17) and dealing with the integration in two parts, thus

$$\int_0^L v \, dx = \int_0^{kL} v \, dx + \int_{kL}^L v \, dx,$$

with the appropriate values of C_1, C_2, C_1', and C_2', we have

$$h = \frac{pkL}{D}\left\{ \frac{kL^2}{12}(3 - 2k) - \frac{1}{c^2} \right\} - \frac{p}{Dc^3(e^{cL} - e^{-cL})}\left\{ e^{cL(1-k)} \right.$$
$$\left. + e^{-cL(1-k)} - e^{cL} - e^{-cL} - e^{kcL} - e^{-kcL} + 2 \right\}, \quad (23)$$

where the denominator D is

$$D = \frac{16d}{c^3L^2} \cdot \frac{e^{cL} - 1}{e^{cL} + 1} - \frac{8d}{c^2L} + \frac{2}{3}\, dL. \qquad (24)$$

Equations (23) and (24) provide direct, but lengthy, means for the evaluation of h due to p. If greater accuracy is required, then (16) must be used instead of (17), and h determined by a trial and error, or graphical process, for any particular value of p.

7.5. It is of interest to study, in terms of the above results, the internal actions on components of the bridge. By inserting (18) in (3), we have for the bending moment on the girder

$$M = h\left\{ C_1 e^{cx} + C_2 e^{-cx} + \frac{1}{c^2}\left(\frac{8d}{L^2} - \frac{p}{h} \right) \right\}. \qquad (25)$$

This, with the appropriate values of the constants C_1 and C_2 from equations (19) to (22), gives the bending moment on any section of the stiffening girder.

By differentiating (25) once we can determine the shear forces on the girder, and by repeating the process we can determine the loading carried by the girder itself. Thus

$$\frac{\partial^2 M}{\partial x^2} = c^2 h\left\{ C_1 e^{-cx} + C_2 e^{-cx} \right\}. \qquad (26)$$

The loading q carried by the suspension rods will be the difference between p and the loading (26); hence

$$q = p - c^2 h\left\{ C_1 e^{cx} + C_2 e^{-cx} \right\}. \qquad (27)$$

We see at once that in general q will vary with x and not be constant, as was assumed in the elastic theory.

7.6. In the foregoing paragraphs 7.4 and 7.5, we have developed the theory for a particular condition of loading. Owing to the non-linear characteristics of the theory, this particularisation is essential, but for practical convenience a large number of cases of loading have in fact been solved in the above manner. Collections of the solutions so obtained are given by Johnson, Bryan and Turneaure[38] (pp. 290–93) and by Steinman[11] (pp. 264–67 of the 2nd edition). In these collections, too, the effects of side spans are introduced.

There follows in the three succeeding chapters some account of ways of simplifying this deflection theory, firstly by linearising its equations, secondly by the use of Fourier series and successive approximation methods, and lastly by approximate empirical methods convenient for preliminary design.

The now general availability of large digital computers has lent a new emphasis and power to some of these approaches, as well as provided a tool for applying the deflection theory outlined in this chapter to cases too complex for direct analytical treatment. The most natural development, typified by the work of D. M. Brotton,[39] has derived from the flexibility coefficient method of paragraph 8.4. By this means the general equations of the deflection theory are replaced by a large number of equations which express in matrix form the matching of the deflections of cable and stiffening girder at every hanger, including provision for hanger extension. In the most general case—very necessary where inclined hangers as in the Severn Bridge are used—these deflections will include horizontal as well as vertical ones. This procedure results in a matrix statement of the suspension bridge problem that is convenient for solution by successive computation methods.

But the development of computers has also, of course, speeded up the relaxation process of paragraph 9.4, and this is still favoured by suspension bridge designers in Great Britain. In their hands, full allowance for horizontal deflections has been included where necessary in the relaxation process.

It is clearly possible in the relaxation method, and inherently in the flexibility coefficient method, to reduce the computational work by starting from the results of an approximate solution so as to reduce the number of successive calculations necessary to reach sufficiently accurate results. For this purpose, the approximate methods of Chapter 10 can provide a basis not only for preliminary design but also for starting more accurate investigations.

CHAPTER 8

The Linearised Deflection Theory

8.1. Attention has been drawn in the foregoing chapter to the non-linear character of the deflection theory, and to some of the difficulties of numerical application resulting therefrom. It was natural therefore that early in the development of the deflection theory attempts should be made to simplify it. In 1894 Godard[40] proposed a linearisation of the theory both for its simplification and for the advantages that it gave by making the use of superposition and influence lines legitimate. With more experience of the deflection theory and its application to bridges of longer and longer spans, the linearised theory became increasingly valuable and relevant. In 1935 H. Bleich[17] outlined the theory in some detail and used it for determining distributions of loading of special design significance. F. Bleich[41] followed this in 1950 with a restatement of the theory, which he used for discussing the natural frequencies and modes of vibration of suspension bridges. At the same time he drew attention, by a diagram of the type shown in Fig. 22, to the essential relations between the elastic, deflection and linearised deflection theories.

It is clear that, provided the live loading p is small compared with the dead loading w, the deflections v will be proportional to p. It is physically clear also that as p increases compared with w, the bridge, due to the action of the cable, will become stiffer. These

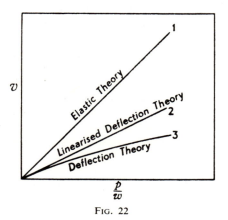

Fig. 22

characteristics are illustrated by curve 3 in the diagram. The linearised theory aims to produce curve 2 (a straight line) tangential to curve 3 at the origin. It is thus accurate for small values of p/w, and even for large values is more accurate than the elastic theory (curve 1). In very long span bridges, the live loading is commonly small compared with the dead loading (i.e. p/w is small) and the linearised theory is thus of particular relevance.

8.2. The fundamental equation of the deflection theory, given as (10) in para. 7.2, and repeated for convenience in (1) below, is a non-linear one

$$EI \cdot \frac{\mathrm{d}^4v}{\mathrm{d}x^4} - (H + h)\frac{\mathrm{d}^2v}{\mathrm{d}x^2} = p + h \cdot \frac{\mathrm{d}^2y}{\mathrm{d}x^2}. \tag{1}$$

So long as the loading p in this equation is small compared with the dead loading w, the horizontal tension h due to p will be small compared with the initial tension H. For such conditions, therefore, (1) can be simplified by the omission of h in the second term, to read thus:

$$EI \cdot \frac{\mathrm{d}^4v}{\mathrm{d}x^4} - H \cdot \frac{\mathrm{d}^2v}{\mathrm{d}x^2} = p + h \cdot \frac{\mathrm{d}^2y}{\mathrm{d}x^2}. \tag{2}$$

The basic equation thus becomes a linear differential equation with h directly proportional to p. As a result, v is also proportional to p, and the principle of superposition and the method of influence lines become applicable.

The second fundamental equation of the deflection theory—(16) of para. 7.3—relating to the extension of the cable still holds and is repeated here as (3) below:

$$\Delta l = \frac{hl}{A_c E_c} - \frac{w}{H}\int_0^L v \,\mathrm{d}x = 0. \tag{3}$$

The linearised theory proceeds by the simultaneous solution of (2) and (3) above, for v and h.

8.3. Tie Analogy Method

There are a number of useful ways in which the theory can now be developed. Some of these ways are not, of course, restricted to the linearised theory, but they are specially useful with it. The first method of analysis given here is of that character.

In equation (2), the term $H \cdot \mathrm{d}^2v/\mathrm{d}x^2$ represents the loading effects of the pull in the suspension rods, and is the same as if the horizontal pull H were applied along the stiffening girder and operating alone on the deflections v. This thought leads to an interesting analogy between the suspension bridge and the more familiar case of a tie rod under lateral loading. If we consider such a tie rod under an axial pull X and lateral loading p, we have as the equation of moments at any section x from one end:

$$EI \cdot \frac{d^2y}{dx^2} - Xy = -\frac{p}{2}\left(\frac{L^2}{4} - x^2\right). \tag{4}$$

Differentiating this twice gives

$$EI \cdot \frac{d^4y}{dx^2} - X \cdot \frac{d^2y}{dx^2} = p, \tag{5}$$

which has the same form as (2). To complete the analogy we must add the lateral loading represented by the last term of (2), which, from (8) of para. 7.2, can be rewritten as $-w \cdot (h/H)$. For a tie rod under an end pull H with this lateral loading present as well as p, (5) becomes

$$EI \cdot \frac{d^4y}{dx^2} - X \cdot \frac{d^2y}{dx^2} = p - w\frac{h}{H}, \tag{6}$$

and is comparable in form with (2) rewritten as

$$EI \cdot \frac{d^4v}{dx^4} - H \cdot \frac{d^2v}{dx^2} = p - w\frac{h}{H}. \tag{7}$$

The well-known solutions of (6) are thus directly available for our problem in the form (7).

As an example, consider the case of a suspension bridge with a single applied load P at a distance x_1 from the end R, as in Fig. 23.

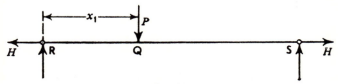

FIG. 23

The effect of the cable will be represented by the tensions H shown together with an upward loading $w(h/H)$ distributed along the whole span. For this case existing solutions give v in terms of hyperbolic functions in which the expression H/EI recurs. Writing

$$\beta^2 = \frac{H}{EI}, \tag{8}$$

these solutions are conveniently dealt with in two parts.

Due directly to the load P, for the region RQ,

$$v' = -\frac{P}{H} \cdot \frac{\sinh \beta(L - x_1)}{\beta \sinh \beta L} \sinh \beta x + \frac{Px(L - x_1)}{HL}, \qquad (9)$$

and for the region QS,

$$v' = -\frac{P}{H} \frac{\sinh \beta x_1}{\beta \sinh \beta L} \sinh \beta(L - x) + \frac{Px_1(L - x)}{HL}. \qquad (10)$$

Associated with the upward pull are the deflections

$$v'' = -\frac{h}{H} \cdot \frac{wL^2}{H} \cdot \left\{ \frac{\cosh(\frac{1}{2}\beta L - \beta x)}{\beta^2 L^2 \cosh \frac{1}{2}\beta L} - \frac{1}{\beta^2 L^2} + \frac{x(L - x)}{2L^2} \right\}. \qquad (11)$$

The total deflection of the bridge girder at any point is thus given from (9) or (10) and (11) by

$$v = v' + v'', \qquad (12)$$

using the appropriate value of v'.

The deflections thus determined are still in terms of the unknown h, and for this we must turn to our second basic equation (3). Substituting our result (12) into equation (3) gives the somewhat lengthy relation between h and the applied load P that is required:

$$h \left\{ \frac{H}{A_c E_c} \frac{l}{L} + \frac{1}{12} \left(\frac{8d}{L} \right)^2 \left(1 - \frac{12}{\beta^2 L^2} + \frac{24}{\beta^3 L^3} \tanh \frac{\beta L}{2} \right) \right\}$$

$$= P \frac{8d}{L} \left\{ \frac{x_1(L - x_1)}{2L^2} - \frac{1}{\beta^2 L^2 \sinh \beta L} [\sinh \beta L - \sinh \beta(L - x_1) \right.$$

$$\left. - \sinh \beta x_1] \right\}. \qquad (13)$$

As pointed out earlier, by the linearised nature of our solution h is directly proportional to P.

Equation (13) can be greatly simplified for long span bridges. In such cases βL is usually large and of the order of 20 or 30, and all the terms containing βL in (13) are so small as to be negligible. Moreover, as pointed out previously, the influence of the cable extension, represented in (13) by $(H/A_c E_c) (l/L)$, is usually very small. If this also is neglected, (13) reduces at once to

$$h = \tfrac{3}{4} P \frac{L}{d} \frac{x_1(L - x_1)}{L^2}. \qquad (14)$$

By writing $x_1 = rL$ as in Chapter 3, and noting that $H = wL^2/8d$ there, (14) is seen to be identical with (15) of para. 3.2. In other words, h as given by (14) is the same as for an unstiffened cable under the load P. This is a valuable approximation for practical use.

Equation (13), or more simply (14), can be used to draw an influence line for h for various positions of the load P. This is done in Fig. 24, which also shows influence lines for the deflection and

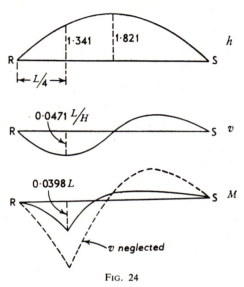

Fig. 24

bending moment at the quarter point C (i.e. $x_1 = L/4$). The former is based upon equations (9), (10) and (11), using as an example the values $\beta L = 10$, $H/A_c E_c = 0.002$, and $L/d = 10$. The latter relates to the same example, using the relation

$$M = \mu + hy + Hv, \tag{15}$$

which is based on (3) of para. 7.1, linearised by the omission of h from the last term $(H + h)v$. The dotted curve in the influence line diagram for bending moment corresponds to (15) with the term Hv omitted; it thus corresponds to the complete neglect of the deflections v and shows the importance of these deflections as affecting the bending moments in a fairly long span bridge.

8.4. Flexibility Coefficient Method

A second way of treating the linearised problem is the flexibility coefficient approach due to Pugsley,[18] a method since developed by T. J. Poskitt[42] to cover the non-linear range. We have already seen in para. 3.3 (c) how the deflections of an isolated cable due to small loads can be considered in terms of a table of flexibility coefficients corresponding to the effects of unit load placed successively at a number of stations across the cable span; and in Table II of para. 3.3, p. 33, a set of such coefficients is given for the case when the unit load employed is small compared with the weight of the cable.

This approach can be extended to provide the basis for an analysis of the linearised suspension bridge problem. Consider first the behaviour of the cable itself and let us treat it as having the whole dead weight of the bridge, including its own weight, so attached to it that each element of weight comprised between two closely spaced transverse vertical sections through the bridge partakes fully of any displacement, vertical or horizontal, of the element of cable contained between the sections. This simplification and idealisation of the behaviour of the elements of weight of the bridge assumes, *inter alia*, that the vertical suspension rods do not extend much, and will naturally be the more accurate the more the weight of the bridge is concentrated in its cables—as tends to happen in bridges of long span.

By applying a small unit load at each of a number of stations across the cable 1, 2, 3, . . . separately in succession, and in each case measuring the deflections at all the stations, we could construct a flexibility coefficient table of the type shown below:

TABLE VI

CABLE COEFFICIENTS

Loaded station	Deflection measured at station number			
	1	2	3	—
1	v_{11}	v_{12}	v_{13}	—
2	v_{21}	v_{22}	v_{23}	—
3	v_{31}	v_{32}	v_{33}	—
—	—	—	—	—
—	—	—	—	—

Here v_{ab} is the deflection at station b due to unit load at station a, and could either be calculated in the manner of Chapter 3 or measured on an appropriate model. To provide complete information regarding the behaviour of the cable, four such tables would be required, one pair referring to the vertical and horizontal deflections due to vertical loads, and the other pair referring to the vertical and horizontal deflections due to horizontal loads. In practice, horizontal movements and loading actions are not often important, and one table like Table VI will suffice.* The adoption of about nine stations evenly spaced across the cable span, corresponding

---

* With the use of inclined hangers, as in the Severn Bridge, horizontal loads and deflections become significant, and all four tables would be needed.

to the mid-points of ten divisions of the span, will commonly provide an adequate knowledge of the cable displacements under load.

There is, of course, no need to restrict our table or tables to the idealised gravity stiffnesses of the cable. We could make our coefficients cover at one and the same time deflections associated with the actual dispositions of weights between cable and deck and those associated with elastic extensions of the cable itself. If such inclusive coefficients proved hard to calculate, we could always resort to measurements on a model that represented the cable elasticity as well as the true gravity system.

Having such tables as Table VI to represent the deflection behaviour of the cable, we can proceed to discuss the behaviour of the stiffening girder on the same lines. Suppose we have adopted nine stations across the cable span and that the girder is simply supported at each end. In the case of a girder that is a framework it may be convenient to choose the stations at joints in the framework. At each station across the girder, treated as an isolated structure with simply supported ends, we could apply in turn a unit vertical load and measure in each case the vertical deflections at all the stations. From the results of this process, whether calculated or measured—and we could include the deflection effects of shearing as well as bending actions—we could construct a flexibility coefficient table of the type of Table VII.

TABLE VII

GIRDER COEFFICIENTS

Loaded station	Deflection measured at station number			
	1	2	3	–
1	V_{11}	V_{12}	V_{13}	–
2	V_{21}	V_{22}	V_{23}	–
3	V_{31}	V_{32}	V_{33}	–
–	–	–	–	–
–	–	–	–	–

Here V_{ab} is the deflection at station b due to unit load at station a.

With these two tables, Table VI and Table VII, we can now approach the problem of the complete bridge. Taking the simple case of a single span with vertical suspension rods, and treating these as long enough to make unimportant any disparities between the small horizontal movements of corresponding points on cable and girder, we can use our table to express the essential compatibility of the vertical deflections of the cable and girder at every station. In doing this we can, of course, allow for the extensibilities of the suspension rods by replacing the actual rods by equivalent rods at each of the chosen stations. By "equivalent" is here meant equality in extensibility to that of the actual rods in the region of the station concerned (i.e. over a distance equal to the spanwise interval between adjacent stations). Let $\alpha_1, \alpha_2, \alpha_3, \ldots$ be the extensibilities of these equivalent rods at the stations 1, 2, 3, . . . respectively; thus unit load in the rod at station 1 will produce an overall extension α_1 of the rod, and so on.

Consider now the effect of, say, a single vertical load W applied on the bridge at station 2. Let the effect of this be to produce tensions $T_1, T_2, T_3, \ldots$ in the equivalent rods. Then the resulting downward vertical displacement of station 1 on the cable will be, from Table VI,

$$T_1 v_{11} + T_2 v_{21} + T_3 v_{31} \ldots = \Delta_{C1}$$

and that of the girder will be, from Table VII,

$$- T_1 V_{11} + (W - T_2)V_{21} - T_3 V_{31} \ldots = \Delta_{G1}$$

The difference $(\Delta_{C1} - \Delta_{C1})$ will be the extension of rod 1, given by $T_1\alpha_1$. Thus we have as the equation of compatibility between cable and girder deflections at station 1,

$$- T_1 V_{11} + (W - T_2)V_{21} - T_3 V_{31} \ldots - T v_{11} - T_2 V_{21} - T_{31}$$

$$\ldots = T_1\alpha_1. \tag{16}$$

Similar equations can be constructed for each of the stations (nine, for example) across the span; at station 2, for example, we have

$$- T_1 V_{12} + (W - T_2)V_{22} - T_3 V_{32} \ldots - T_1 v_{12} - T_2 v_{22} - T_3 v_{32}$$

$$\ldots = T_2\alpha_2, \tag{17}$$

and so on. We thus have a set of simultaneous linear algebraic equations with as many unknowns—T_1, T_2, T_3, etc.—as equations. The tensions T_1, T_2, T_3, can thus be solved directly in terms of the applied load W. Because the flexibility coefficients have been assumed constant, corresponding to small deflections, these tensions will be directly proportional to W.

Once the tensions T_1, T_2, T_3, etc. are known, the vertical deflections at each station of the girder can be calculated from expressions such as that for Δ_{G1} given above. The shears and bending moments in the girder itself can, of course, be evaluated at once by considering the girder as an isolated beam under the joint action of the forces T_1, T_2, T_3, together with W and any end reactions necessary to equilibrate these forces.

8.5. Energy Method

A third way of treating the linearised problem is by means of energy relations, using trigonometrical series for the approximate representation of the deflections. F. Bleich[41] used this procedure, but the series method was, as in so many problems, initiated by Timoshenko (1930)[14].

A neat statement of this procedure, cast in terms of either flexibility coefficients or of stiffness coefficients at a finite number of stations along the bridge span (much as used in the foregoing section 8.4), has recently been given by C. F. P. Bowen and T. M. Charlton.[43] They state the procedure in two forms: one in terms of complementary energy, leading to compatibility of deflection equations in terms of the hanger tensions T_1, T_2 . . . as in the flexibility approach just outlined; and another in terms of total potential energy, leading to equilibrium equations giving the deflections v_1, v_2 . . . at the hanger stations. They find that the latter procedure has some advantage in that, for the same accuracy, fewer terms are required in the Fourier series used. We will confine our attention here to the total potential energy method.

We consider a single span of length L with a number n of equally spaced stations 1, 2, 3 . . . n, along it (in 8.4 we adopted nine such stations). Due to a system of superimposed loads W_1, W_2 . . . applied at these stations, let the deflection of the girder be represented by the finite series with p terms

$$v = \sum_{k=1}^{k=p} B_k \sin \frac{k\pi x}{L}. \tag{18}$$

In these circumstances, the bending strain energy stored in the girder (assumed to be of uniform stiffness) will be

$$U_1 = \tfrac{1}{2}EI \int_0^L \frac{\mathrm{d}^2 v}{\mathrm{d}x^2}\, \mathrm{d}x. \tag{19}$$

Now the linear stiffnesses of the cable system can be represented by a matrix of stiffness coefficients, the inverse of the matrix (represented by Table VI) of flexibility coefficients discussed in section 8.4 above. A typical stiffness coefficient b_{ij}, where i and j are stations along the span, will correspond to the vertical force at station i that is necessary to prevent any vertical deflection there when unit vertical deflection is given to the system at station j, and all other stations 1, 2, 3 . . . n along the span are also prevented from deflecting. Thus for the vertical deflections v_1, v_2 . . . v_n at the stations the corresponding vertical forces are

$$F_1 = b_{11}v_1 + b_{12}v_2 + \ldots + b_{1n}v_n,$$
$$F_2 = b_{21}v_1 + b_{22}v_2 + \ldots + b_{2n}v_n, \text{ etc.}$$

and the potential energy stored in the deflected cable system is

$$U_2 = \tfrac{1}{2} \sum_{i=1}^{j=n} \sum_{i=1}^{i=n} b_{ij} v_i v_j. \tag{20}$$

In addition, the deflection (18) "lowers" the applied forces W_1, W_2..., so that a loss of potential energy occurs equal to

$$U_3 = \sum_{i=1}^{i=n} W_i v_i. \tag{21}$$

The net change of potential energy is thus

$$V = U_1 + U_2 - U_3, \tag{22}$$

which, by substituting from (18) for v_1, v_2 . . . v_n, can be expressed in terms of B_1, B_2 . . . B_p.

For equilibrium this energy change must be stationary with respect to the deflection parameters B_k. Hence

$$\frac{\partial V}{\partial B_1} = \frac{\partial V}{\partial B_2} = \ldots \frac{\partial V}{\partial B_p} = 0, \tag{23}$$

giving as many simultaneous linear equations as the unknowns B_k.

The convenience of this statement of the method is that, whereas in the full flexibility coefficient approach of 8.4 the number of simultaneous equations to be solved is equal to the number of equivalent hanger stations, in this case the number of unknowns B_k to reach a good approximation is found to be only four or five. It is relatively easy therefore, to solve (23) for B_1, B_2, B_3 . . ..

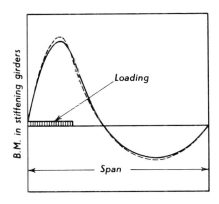

FIG. 25

It is important to note that, when the bending moments M in the stiffening girder are required, it is unwise to try to deduce them by double differentiation for $\dfrac{d^2v}{dx^2}$ and equating this to $\dfrac{M}{EI}$. Much more accurate results are obtained by calculating the hanger tensions T from the deflections at the n stations given by (18), and the conditions for local equilibrium, giving

$$T_i = W_i - F_i, \tag{24}$$

and thence calculating the bending moments at these stations directly from the equilibrium conditions, using the forces $T_1, T_2 \ldots$ and the loads $W_1, W_2 \ldots$.

Fig. 25, based upon the results of Bowen and Charlton, illustrates the application of this method to a single span bridge with uniform loading spread along one quarter of its span from one end. The continuous line shows the girder bending moments as given by the full deflection theory, and the dotted line shows the results of the above energy method, using only four parameters B_k.

CHAPTER 9

The Fourier Series Treatment of the Deflection Theory

9.1. Reference has already been made, in para. 8.5, to the use of trigonometrical series for the treatment of the linearised deflection theory. A series approach can be employed for the deflection theory in its general form, and is indeed the procedure most commonly employed today by suspension bridge designers.

In 1930, a solution of the deflection theory equations by expressing the deflections v in terms of a trigonometrical series was given in a paper (1930) by S. Timoshenko[14] and further developed in the discussion of the paper, particularly by E. Steuerman. In this solution the series method was employed in an energy presentation of the problem, and the non-linearity of the full equations employed was treated by an approximate trial and error method for the tension h. and error method for the tension h.

In 1938, Southwell, then in course of developing his Relaxation Method,[44] considered the suspension bridge problem as an example of a non-linear structural problem that might conveniently yield to the method. He and Atkinson[15] developed a procedure on these lines, taking the opportunity at the same time to demonstrate how the effects of horizontal movements of the cable could be studied. In this work the problem was reduced to relating, by the relaxation process, the coefficients in two Fourier series, one for v and the other for the simple bending moment μ defined by

$$\frac{\mathrm{d}^2\mu}{\mathrm{d}x^2} + p = 0, \tag{1}$$

where $\mu = 0$ at $x = 0$ and $x = L$.

This was followed, during the recent war, by a further paper from Timoshenko[45] which not only produced a much developed version of his original series method, but also discussed the effects of various commonly neglected terms, such as the horizontal cable movement just mentioned, in a general physical way that has provided an excellent basis for examining *ab initio* their practical effects. In this version of his solution, the treatment of the essential non-linearity of the problem was systematised by a simple interpolation procedure for the evaluation of the additional cable tension h.

Following upon this work, C. D. Crosthwaite[46] in the course of his design studies with Messrs. Freeman, Fox & Partners for a proposed Severn Bridge, showed how the relaxation method could be applied to a practical design of bridge, making allowance for the effects of suspension rod extension and girder non-uniformity as well as for horizontal movements of the cable.

These modern developments all hinge around the use of Fourier series, and in their presentation—necessarily in brief only here— it has seemed desirable to outline the series solution first and thereafter comment on the associated use of the relaxation process.

9.2. Series Method

It will be convenient to outline the series method in terms of the case of a single span bridge under a concentrated load P at $x = x_1$. The fundamental differential equation of the deflection theory becomes for this case, from (11) of para. 7.2,

$$EI \frac{d^2v}{dx^2} - (H + h)v = h \frac{4d}{L^2} x(L - x) - \frac{Px(L - x_1)}{L}, \qquad (2)$$

for $x < x_1$, and

$$EI \frac{d^2v}{dx^2} - (H + h)v = h \frac{4d}{L^2} x(L - x) - \frac{Px_1(L - x)}{L}, \qquad (3)$$

for $x > x_1$.

Now for the whole span L, the right-hand sides of (2) and (3) can be expressed in terms of one Fourier series, such as

$$b_1 \sin \frac{\pi x}{L} + b_2 \sin \frac{2\pi x}{L} + b_3 \sin \frac{3\pi x}{L} + \ldots, \qquad (4)$$

where a typical coefficient b_m must be such that*

$$
\begin{aligned}
b_m = \frac{2}{L} \Bigg\{ & h \frac{4d}{L^2} \int_0^L x(L - x) \sin \frac{m\pi x}{L} \cdot dx \\
& - \frac{P(L - x_1)}{L} \int_0^{x_1} x \sin \frac{m\pi x}{L}\, dx - \frac{Px_1}{L} \int_{x_1}^L (L - x) \sin \frac{m\pi x}{L}\, dx. \Bigg\}
\end{aligned}
\tag{5}
$$

On integration, this equation (5) becomes

$$b_m = \frac{16hd}{m^3\pi^3} (1 - \cos m\pi) - \frac{2PL}{m^2\pi^2} \sin \frac{m\pi x_1}{L}. \qquad (6)$$

---

* This equality is obtained in the standard manner by equating the R.H.S. of (2) or (3) to (4), multiplying both by $\sin m\pi x/L$ and integrating between the given limits.

We can now rewrite the fundamental equations (2) and (3) in terms of b_m as the one equation

$$EI\frac{d^2v}{dx^2} - (H - h)v = \sum_{m=1}^{m=\infty} b_m \frac{\sin m\pi x}{L}. \tag{7}$$

Let us assume as the solution of (7) a Fourier series for the deflection v thus

$$v = \sum_{m=1}^{m=\infty} a_m \sin \frac{m\pi x}{L}. \tag{8}$$

By substituting this in (7) and equating the coefficients of the resulting two series, one on each side of the equation, we have

$$a_n = -\frac{b_m L^2}{EIm^2\pi^2 + (H + h)L^2}, \tag{9}$$

whence, from (8),

$$v = -\sum_{m=1}^{m=\infty} \frac{b_m L^2 \sin \dfrac{m\pi x}{L}}{EI(m^2\pi^2 + k^2L^2)}, \tag{10}$$

where $k^2 = (H + h)/EI$. But we have already evaluated b_m in (6), so that (10) can now be written in terms of the physical parameters of the problem thus:

$$v = \frac{32hdL^2}{\pi^3 EI} \sum_{1}^{\infty} \frac{\sin \dfrac{m\pi x}{L}}{m^3(m^2\pi^2 + k^2L^2)} + \frac{2PL^3}{EI\pi^2} \sum_{1}^{\infty} \frac{\sin \dfrac{m\pi x_1}{L} \sin \dfrac{m\pi x}{L}}{m^2(m^2\pi^2 + k^2L^2)}. \tag{11}$$

It remains to determine the value of the additional tension h due to the load P. We have already seen, in para. 8.3, how this can be done directly in the case of the linearised deflection theory, giving equation (13) of para. 8.3 as a result. If we follow the same procedure again, but do not linearise the problem by neglecting h compared with H, we find, instead of (13) in para. 8.3, the very similar relation

$$h\left\{\frac{H + h}{A_cE_c}\frac{l}{L} + \frac{1}{12}\left(\frac{8d}{L}\right)^2\left(1 - \frac{12}{k^2L^2} + \frac{24}{k^3L^3}\tanh\frac{kL}{2}\right)\right\}$$

$$= P\frac{8d}{L}\left\{\frac{x_1(L - x_1)}{2L^2} - \frac{1}{k^2L^2\sinh kL}[\sinh kL - \sinh k(L - x_1)\right.$$

$$\left. - \sinh kx_1]\right\}. \tag{12}$$

As before, for long span bridges, this breaks down as an approximation to the form

$$h = \tfrac{3}{4}P\frac{L}{d}\frac{x_1(L - x_1)}{L^2} \tag{13}$$

appropriate to an unstiffened cable. In his original paper, Timoshenko suggested for the solution of the non-linear equation (12)
the following procedure. From (13), calculate a first value of h
and thence determine k. With this value of k, find a second (and
smaller) value of h from (12). This will usually give a sufficiently
good approximation without further successive calculations.

Another and more consistent procedure for determining h is given
in the 1943 paper. We can substitute the series (8) for v in the
fundamental equation for the length of the cable

$$\frac{hl}{AE} - \frac{w}{H} \int_0^L v \, \mathrm{d}x = 0$$

from (16) of para. 7.3. If we do this, we obtain as a direct result
for h,

$$\frac{hl}{A_c E_c} = \frac{w}{H} \frac{2L}{\pi} \left(a_1 + \frac{a_2}{3} + \frac{a_3}{5} + \cdots \right)$$
$$+ \frac{\pi^2}{4L} (a_1{}^2 + 2^2 a_2{}^2 + 3^2 a_3{}^2 + \cdots). \quad (14)$$

This gives rise to the following procedure. As before, first determine
h from (13), and thence k. With these values of h and k, calculate
$b_1, b_2, b_3, \ldots$ from (6) and thence $a_1, a_2, a_3, \ldots$ from (9). Substituting these values for a_1, a_2, a_3 in (14) gives a second approximation
to h. If it were necessary to proceed further, we could now repeat
the whole process, starting with a new value of k.

The foregoing discussion relates to the case of a single load P;
Timoshenko has also dealt similarly with a short distributed load.
For a distributed load of the order of half the span in length, the
second group of terms in (14) prove to be negligible compared with
the first; this can be shown to arise because this second group
relates to the degree of non-uniformity of the suspension rod loads
across the span, which is obviously most marked for a concentrated
load.

9.3. Throughout the foregoing analysis the stiffening girder has
been treated as of uniform section. The problem of dealing with
the case of a girder with varying section was not solved in terms of
the deflection theory until 1930, when Steuerman treated it by an
extension of Timoshenko's series method, as follows.

Let the variation of EI be expressed by the equation

$$EI = EI_0 \phi(x), \quad (15)$$

where $\phi(x)$ is a known function of x. Looking back at the solution
of para. 9.2 with this in mind, we see that no change in procedure
is required till we reach equations (7) and (8). Inserting (15) in (7),

we find then that instead of the simple relation between a_m and b_m expressed by (9), we have

$$- \frac{\pi^2}{L^2} EI_0 \phi(x) \sum_1^\infty m^2 a_m \sin \frac{m\pi x}{L} - (H + h) \sum_1^\infty a_m \sin \frac{m\pi x}{L}$$

$$= \sum_1^\infty b_m \sin \frac{m\pi x}{L}. \quad (16)$$

To render this equation numerically tractable, let us multiply both sides by $\sin i\pi x/L$, having in mind values of i of 1, 2, 3, . . . up to a finite number n. Then integrating the resulting equation we have

$$b_i = - \frac{2\pi^2}{L^3} EI_0 \sum_1^\infty m^2 a_m \int_0^L \phi(x) \sin \frac{m\pi x}{L} \sin \frac{i\pi x}{L} \, dx - a_i(H + h). \quad (17)$$

For a given function $\phi(x)$, we can evaluate the integral in (17) and obtain n linear equations for $a_1, a_2, a_3, \ldots a_n$ in terms of $b_1, b_2, b_3, \ldots b_n$. The latter can be determined from (6) for an assumed value of h, and $a_1, a_2, a_3, \ldots a_n$ so calculated. A check on the assumed value of h can then, of course, be made from (14).

As a result of calculations of this kind, Timoshenko found that h is not very sensitive to practical variations of EI and that a good approximation to h can be obtained by using an average EI value and treating the girder as uniform.

9.4. Relaxation Treatment

When Southwell's relaxation method was first applied to suspension bridge problems, particular interest was attached to the influence of horizontal movements of the cable and to the case of a variable section stiffening girder. The former influence, partly as a result of the relaxation work, is now known to be small in long span bridges with vertical hangers, but the presence of variable section girders is still a reason for the use of relaxation methods in the suspension bridge field; in such a case, as our brief examination of the matter in para. 9.3. above shows, Steuerman's extension of the series method is by no means short or simple. Moreover, where inclined hangers are adopted, some treatment of the horizontal actions must always be introduced.

The relaxation method was described in some detail by Atkinson and Southwell[15] and again with some extensions and corrections by Crosthwaite.[46] Here we will endeavour only to outline the procedure in relation to the simple single span problem. The method starts by expressing both the vertical deflections v and the equivalent simple beam bending moments in terms of Fourier series. Thus, we write, as in (8)

$$v = \Sigma a_m \sin \frac{m\pi x}{L} \qquad (18)$$

and for μ we take

$$\mu = \Sigma c_\lambda \sin \frac{\lambda \pi x}{L}, \qquad (19)$$

where the symbols m and λ both refer to the series 1, 2, 3, . . . taken to some convenient finite number—9 was adopted by South-well in his first example, and 16 by Crosthwaite.

Now the coefficients c_λ are known, or determinable, for any given external loading on the bridge; the series (19) has, in fact, to be so chosen as to represent adequately the simple bending moments μ everywhere across the span. We have, therefore, to find the deflection coefficients a_m in terms of the known coefficients c_λ. To do this, we resort to the basic equations for the bridge, such as (10) of para. 7.2 and (16) of para. 7.3. By introducing (18) and (19) into these two fundamental equations, we can rewrite them in the form

$$c_\lambda - h \sum_m (A_{\lambda,m} + B_{\lambda,m})a_m = H \sum_m F_{\lambda,m}a_m, \qquad (20)$$

$$h = \sum_m D_m a_m. \qquad (21)$$

Here $A_{\lambda,m}$, $B_{\lambda,m}$, D_m, $F_{\lambda m}$ are all quantities that are fixed for a given design of bridge, and can be tabulated at the outset for various values of λ and m. They will, incidentally, include allowance for any variation in EI along the girder; this is done by writing, in the first fundamental equation, the term $EI/(\mathrm{d}^4v/\mathrm{d}x^4)$ as $(\mathrm{d}^2/\mathrm{d}x^2)(EI(\mathrm{d}^2v/\mathrm{d}x^2))$.

Equations (20) and (21) form the basis for the relaxation calculations. To allow for their essential non-linearity, h is first neglected in both equations, which then reduce to a series of linear equations relating the coefficients a_m and c_λ, and these are solved by ordinary relaxation methods. We thus derive a first approximation to the coefficients a_m. With these first values, (21) then gives a first value for h, and with this the whole procedure can be repeated, and so on till h settles to an almost constant value. The corresponding coefficients a_m then, by (18), give the deflection v everywhere, and any further information required, such as the bending moments on the girder, can be deduced therefrom.

CHAPTER 10

Approximate Methods of Analysis
for Preliminary Design

10.1. With the development of the deflection theory, the suspension bridge designer entered a region where the general structural engineer ceased to follow and where methods of analysis became essentially numerical and less capable of use for physical generalisation. As always in such cases—the flutter problem of the aeronautical engineer is an outstanding modern example—a series of efforts have been made to produce simplified versions of the theory that would both emphasise its main physical ingredients and make for rapid computation. By such means bridge engineers have sought, in the first instance, to produce quick methods of analysis for their own preliminary design work, and, as a secondary issue, to provide an introductory approach to their problems for the use of students. It is here proposed to outline, in varied degrees of detail, three such methods of approach.

10.2. Steinman's Modified Elastic Treatment

It is natural, from the history of our subject, that the first method of this kind arose from the elastic theory. For relatively short spans and stiff girders, this theory on its own gives reasonable results;* thus a possible approach was to regard the elastic theory as one easily understood by structural engineers generally and to seek to produce a simple means of modifying its results so that they should agree approximately, for a wide range of bridges, with those of the deflection theory. This was the procedure adopted by Steinman and Baker and outlined by the former in his well-known book.[11]

We have already, in Chapters 5 and 6, described in some detail the elastic theory and its application; Steinman first systematised and simplified its use in design by a series of charts giving girder bending and shears, etc., for a variety of bridge and loading conditions, and then, with Baker, sought for ways in which these conveniently charted results could be *corrected* to approximate to those of the deflection theory. It was realised that the changes, as between

---

* Steinman treats mainly of the theory of long-span suspension bridges, but there are still arguments for the use of short-span suspension bridges in some circumstances —see, for example, *An Economic Analysis of Short-span Suspension Bridges for Modern Highway Loadings*, by McCullough, Paxson, and Smith, Tech. Bulletin No. 11, Oregon State Highway Commission, U.S.A., 1938.

one theory and the other, were due mainly to the terms H and v, and by a consideration* of the physical factors governing these a non-dimensional stiffness parameter

$$S = \frac{1}{L}\sqrt{\frac{EI}{H}} = \sqrt{\frac{8EId}{wL^4}} \qquad (1)$$

was adopted as a factor governing the relation between the results of the elastic and deflection theories.

With this factor as the essential variable, a series of correction curves, of which Fig. 26 is typical, were prepared by comparing the

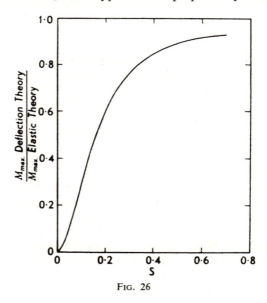

FIG. 26

results of the two theories when applied to a series of bridges having different values of S but the same typical geometrical proportions and other design conditions. Fig. 26 gives the resulting correction curve for the bending moments on the central span of a suspension bridge, and is designed to be applied to the maximum moments (positive and negative) on the girder. Similar curves, all based upon the numerical work of A. H. Baker conducted in the foregoing manner, are given in Steinman's book for shears in the main spans, and for the moments and shears in the side spans of suspension bridges. Supplementary small adjustments for departures from the

---

* This consideration can, of course, be cast in terms of dimensional theory; this is done in para. 10.4.

typical dip to span ratio (0·10), dead loading to live loading ratio ($w/p = 3$), etc., are also given.

10.3. Hardesty and Wessman's Cable Treatment

With the increasing spans of suspension bridges, the stiffening girder has come to play a less important part; in the Washington bridge over the Ohio there was until recently no effective stiffening girder at all. As a result an alternative, and indeed, opposite approach to that initiated by Steinman, has been put forward for approximate analysis work. In 1939, Hardesty and Wessman proposed[20] a method that has since been extensively used which starts, not from the structure viewed as an elastic one, but from the cable viewed on its own as an inextensible one in a gravity field.

The method, with its design aim predominant, starts by studying the lengths of uniform live loading p that give rise to maximum cable deflections at the quarter span points and at the centre. By treating the cable as parabolic in form under a uniform dead loading w, these critical loading conditions are found to be as follows:

TABLE VIII

Condition	Loaded length
¼ span—maximum downward deflection	0·4L at same end of span
¼ span—maximum upward deflection	0·6L at far end of span
Centre—maximum downward deflection	0·3L at centre of span
Centre—maximum upward deflection	0·35L at both ends of span

These results are, of course, generally consistent with those discussed in Chapter 3.

The next step is to assume that the loading that gives these maximum deflections is identical with that giving maximum bending moments at the same points in any stiffening girder present. This the authors support by supplementary evidence from fuller bridge calculations and by considering the form of the girder deflections to be dominated by those natural to the cable. By the same sort of argument, they assume that if a moment M_t is induced in the girder when it is bent to the free cable deflection curve v_t, then the actual moment M in the girder corresponding to the actual deflection v is given by

$$M = M_t \cdot \frac{v}{v_t}. \tag{2}$$

With this approximate relation in mind, and concentrating, as did Steinman, on the influence of H and v—in this case on the appropriate change of moment $(H + h)(v_t - v)$—the authors rewrote (2) in the form

$$M = M_t \cdot \frac{(H + h)v_t}{M_t + (H + h)v_t},\qquad(3)$$

and base their approximate analysis thereon.

We have first to estimate M_t. This will be proportional to v_t for a given distance jL between the nodes (points of zero vertical deflection) in the girder deflection curve, assumed to be governed by the relevant natural mode of the cable. On this basis, we can write

$$M_t = \frac{EI}{R} = C \cdot \frac{EI\, v_t}{(jL)^2},\qquad(4)$$

where C is a constant depending on the precise shape of the deflection curve; for a circular arc C would be 8. For the maximum moments at the quarter span points, Hardesty and Wessman adopt $C = 9\cdot1$ and j about 0·45, and for the centre $C = 9\cdot0$ and j about 0·38. As a result, they give

$$M_t = K \cdot \frac{EI\, v_t}{L^2},\qquad(5)$$

where for the quarter point K is 47·0 for maximum positive moment and 43·0 for maximum negative moment, and for the centre K is 65·8 and 59·2 for the same cases.

The second step in the use of (3), and of (5) therein, is to estimate v_t. This is done by an analysis of the free cable as in Chapter 3, and is simplified by the use of the curves given in Fig. 27, where the values of v_t are given as percentages of the cable dip d for various values of the loading ratio p/w. The required values of v_t are obtained directly from such curves.

It remains to consider the evaluation of $(H + h)$. This problem was discussed in general terms in Chapter 3. The horizontal tension H due to the dead loading w is of course given by

$$H = \frac{wL^2}{8d} = 0\cdot125\,\frac{wL^2}{8d}.$$

The additional tension h is usually smaller than this and is given by

$$h = k \cdot \frac{pL^2}{D},$$

where k varies from 0·040 to 0·085 for the design cases concerned, and D is the appropriate total dip. If we neglect the difference between d and D—seldom more than a few per cent—we can now write

$$H + h = \frac{1}{d}(0 \cdot 125wL^2 + k \cdot pL^2), \qquad (6)$$

where for the quarter point cases k is 0·040 (positive moment) and 0·085 (negative moment) and for the centre cases 0·0638 and 0·0613 respectively.

We thus have, with a fair degree of accuracy, all the terms in (3) from which to estimate design moments at the quarter and centre

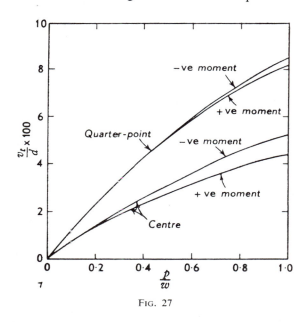

Fig. 27

span points along the stiffening girder. Hardesty and Wessman give various ways of refining the above process, including allowances for the effects of cable extension due to load and temperature changes, and for the effects of side-span interaction.

10.4. Elastic Foundation Analogy

We have seen that the foregoing two methods of approximate analysis arise as extrapolations from two extreme approaches to the suspension bridge problem. In the one elastic actions are assumed to dominate and in the other gravity actions. It is interesting* to look at the general question arising in terms of dimensional theory.

* See, for details, Ref. 22.

The behaviour of a loaded cable, as we have seen in Chapters 2 and 3, depends primarily on its geometry, defined by L and d, and its loading w. That of a girder depends mainly on its span L and its stiffness EI. We may thus reasonably assume that, due to a load P, the moment on a given section of the stiffening girder can be expressed in terms of P, L, d, EI and w. Viewed as a problem in dimensional theory, we can therefore at once write

$$M = PL\psi \left\{ \frac{EId}{wL^4}, \frac{L}{d}, \frac{P}{wL} \right\}. \tag{7}$$

Let us study the parameters in brackets here in turn. The elastic stiffness of the girder, considered alone, when measured by a load W applied at its centre and producing a deflection δ there, is given by

$$\left(\frac{W}{\delta} \right)_E = \frac{48EI}{L^3}. \tag{8}$$

The gravity stiffness of the loaded cable, similarly measured, is, from (19) of para. 3.2, given by

$$\left(\frac{W}{\delta} \right)_G = \frac{2wL}{d}. \tag{9}$$

Thus the ratio of these two stiffnesses is

$$R' = \frac{\text{Elastic stiffness}}{\text{Gravity stiffness}} = 24 . \frac{EId}{wL^4} = 24R. \tag{10}$$

In other words, the first parameter in the brackets of (7) measures the stiffness ratio R. As would be expected, R is also directly related to the stiffness factor S of Steinman and Baker, defined by (1) above.

The second parameter in the brackets of (7), the shape factor L/d is known, within the small variations of L/d adopted in practice, to be unimportant. The third factor, P/wL, can also be written in the alternative form PL^3/EId, expressing the ratio of girder deflection to cable dip. Its omission is clearly consistent with the assumption of small deflections and approximately linear response to load. Thus equation (7) can, for most practical purposes, be reduced to the simple form

$$M = PL \phi \left\{ \frac{EId}{wL^4} \right\}. \tag{11}$$

Of the two approximate theories so far examined, it is clear that the first is most relevant when the stiffness parameter of (11) is high and the second when it is low. The third approximate theory given below has the merit of introducing this stiffness parameter explicitly.

10.5. It is clear from the foregoing dimensional theory that the behaviour of a suspension bridge depends largely upon the ratio of the elastic stiffness of its girder to the gravity stiffness of its cable. Now a relative stiffness problem well known to structural engineers is that of an elastic beam upon an elastic foundation, which has a fully developed literature of its own, and if we can parallel the elastic foundation of this problem with the loaded cable of the suspension bridge one, then the well-known results for the former can be made to apply to the latter.

Now it will have been observed in Fig. 8 of Chapter 3 that the gravity stiffness of a cable is roughly constant for any point along its span except close to its ends. The nine diagonal values of the flexibility table (Table II) in para. 3.3 (*c*) illustrate the same point; their average value is 5·78 per cent for a point load of 0·1*wL*. For an elastic foundation to provide the same stiffness, we can conceive of it as consisting of nine vertical springs, each carrying a load of 0·1*wL* and deflecting 0·0578*d*. The corresponding "modulus of foundation," to use the terminology of elastic foundation literature, would be

$$k = \frac{0 \cdot 9wL}{L \times 0 \cdot 0578d} = 15 \cdot 5 \cdot \frac{w}{d}.$$

This gives at once the form and order of the equivalent modulus* required for our analogy between the cable and an elastic foundation. We will adopt

$$k = \alpha \frac{w}{d}, \tag{12}$$

where α is expected to be of the order of 15, but will best be determined empirically. In the discussion of the original paper putting forward this simple analogy method,[22] some examples quoted by representatives of Messrs. Freeman, Fox and Partners indicated values of from 10 to 12 to give agreement with deflection theory results for moments and deflections in long span bridges. For further information regarding the choice of values for α for design purposes, reference may be made to a more recent paper by the Author.[23]

We can now proceed, with the value of k in (12), to use the existing results for beams on elastic foundations for studying the moments on stiffening girders. Using the collection of these results in Hetenyi's book,[47] we have, as two results of special interest to suspension bridge designers, the following.

* In units corresponding to force per unit length of beam per unit deflection, e.g. tons per foot length per foot deflection.

(a) For a single load P at a distance a from one end $(x = 0)$ of the bridge

$$M = \mu - \frac{2PL}{\pi^2} \sum_{n=1}^{n=\infty} \frac{\sin \dfrac{n\pi a}{L} \cdot \sin \dfrac{n\pi x}{L}}{n^2 \left(1 + n^4 \dfrac{\pi^4}{\alpha} \cdot \dfrac{EId}{wL^4}\right)}, \tag{13}$$

$$v = \frac{2PL^3}{\pi^4 EI} \sum_{n=1}^{n=\infty} \frac{\sin \dfrac{n\pi a}{L} \cdot \sin \dfrac{n\pi x}{L}}{n^4 + \dfrac{\alpha}{\pi^4} \dfrac{wL^4}{EId}}. \tag{14}$$

(b) For a uniform loading p spread over a length c of the span, starting at a distance a from one end $(x = 0)$ of the bridge

$$M = \mu - \frac{2pL^2}{\pi^3} \sum_{n=1}^{n=\infty} \frac{\left[\cos \dfrac{n\pi a}{L} - \cos \dfrac{n\pi(a + c)}{L}\right] \sin \dfrac{n\pi x}{L}}{n^3 \left(1 + n^4 \dfrac{\pi^4}{\alpha} \dfrac{EId}{wL^4}\right)}, \tag{15}$$

$$v = \frac{2pL^4}{\pi^5 EI} \sum_{n=1}^{n=\infty} \frac{\left[\cos \dfrac{n\pi x}{L} - \cos \dfrac{n\pi(a + c)}{L}\right] \sin \dfrac{n\pi x}{L}}{n \left(n^4 + \dfrac{\alpha}{\pi^4} \dfrac{wL^4}{EId}\right)} \tag{16}$$

In these expressions μ stands, as usual, for the bending moment that the external loading would produce on the girder at x were the cable not present.

It will be seen that the expressions (13) and (15) parallel in form the dimensional theory result (11). In practice, the series solutions given converge so rapidly that only the first 4 or 5 terms are significant. The results for M and v are, of course, all linear with the loads P or p. The results for v, incidentally, give directly the distribution of the suspension rod tensions; the foundation reaction per unit length of girder is kv and this corresponds to the rate of rod tension.

It is clear that for a long span bridge with a flexible girder, if the loaded length c in case (b) is small compared with the span and not close to its ends, then the results for M and v in the region of the load will be little affected by the end conditions. In such a case, we may turn to the results for a loaded beam of infinite length on a semi-infinite foundation. In this case

$$M = \frac{p}{4\lambda^2} (e^{-\lambda a} \sin \lambda a + e^{-\lambda b} \sin \lambda b), \tag{17}$$

where the load p is spread over a length $(a + b)$, and M relates to the point dividing the loaded length into the parts a and b. The maximum moment will occur at the centre, where $a = b$, and (17) gives for this

$$M_{max} = \frac{p}{2\lambda^2} \cdot e^{-\lambda a} \sin \lambda a. \tag{18}$$

In both (17) and (18), $\lambda = (k/EI)^{\frac{1}{4}}$, where k is the foundation modulus. From (18), it will be seen that for a given loading intensity p, M_{max} will vary with the semi-length a of the loading. By differentiating (18) we find that M_{max} will be greatest when

$$\lambda a = \frac{\pi}{4}, \tag{19}$$

giving

$$M_{max} = 0 \cdot 161 \frac{p}{\lambda^2}. \tag{20}$$

By substituting from (12) in the expression for λ, these equations become

$$\frac{2a}{L} = \frac{\pi}{4} \left(\frac{4}{\alpha} \cdot \frac{EId}{wL^4} \right)^{\frac{1}{4}}, \tag{21}$$

and

$$M_{max} = 0 \cdot 161 pL^2 \left(\frac{4}{\alpha} \cdot \frac{EId}{wL^4} \right)^{\frac{1}{2}}. \tag{22}$$

Equation (21) illustrates a marked difference between this approximate analysis and that of Hardesty and Wessman; the design loaded length ($2a$ in this special case) is a function (albeit only to the $\frac{1}{4}$ power) of the stiffness ratio EId/wL^4 and not independent of it. This is a qualitative advantage over the approximate methods previously outlined, particularly for bridges with very flexible girders. When the girder is relatively stiff, however, (21) may give a result in excess of the $0.3L$ to $0.6L$ natural to an almost inextensible cable; but in such cases the assumption behind equation (17) that the loaded length $2a$ is small compared with L no longer holds, and resort must be made to equations (15) and (16).

The same point arises when concentrated loads are to be considered, when the deflection effects may be too local to be represented by the earlier approximate theories and yet can be reasonably covered by this foundation analogy approach. In some recent tests[48] on the Clifton Suspension Bridge, which has unusually flexible stiffening girders, it was clear from the suspension rod tension measurements that the concentrated loads used were carried up to the *cables* very locally, and this was deduced also from deflection theory calculations.

In this statement of the Foundation Analogy method, attention has been concentrated upon the girder moments and deflections under, or close to, the loaded region; that is, upon negative bending moments and positive deflections. These are of special design significance, but it should be noted that this method, by its nature, whilst capable of providing good estimates for these quantities, does not usually provide quantitatively good results for any positive bending moments and negative deflections.

CHAPTER 11

The Natural Frequencies and Modes of Suspension Bridges

11.1. Throughout the history of suspension bridges, their tendency to oscillate under dynamic and wind loading has been a matter for remark, and in modern times has led to several investigations into their vibration properties.

The first problem in this field, and a difficult one, concerns the oscillations in a vertical plane of an isolated heavy suspension cable. The earliest known theoretical treatment of this problem was by Rohr in 1851.[19] He examined the symmetrical modes of a cable that is nearly horizontal and produced results for the first two natural modes. The same problem was examined more generally by Routh[50] at the turn of the century, and his treatment merits summarising here as an introduction to the subject.

11.2. Simple Cable

Let x and y be the rectangular coordinates of any point Q on a suspension cable hanging in equilibrium, x being as usual measured horizontally. Let s be the distance along the arc of the cable to Q from some convenient origin, and m the mass per unit length of cable at Q (we envisage here the possibility of m varying along the cable). When the cable is executing a small amplitude oscillation about its equilibrium position, the coordinates of Q, at time t, may be written $x + u$, $y + v$. Let T be the equilibrium tension in the cable at Q and let it become $T + \Delta T$ at time t. By examining the conditions governing the motion of an element of length ds of the cable at Q, Routh states the equations of motion for the cable in the form

$$\frac{\mathrm{d}^2 u}{\mathrm{d}t^2} = \frac{1}{m}\frac{d}{\mathrm{d}s}\left(T\frac{\mathrm{d}u}{\mathrm{d}s} + \Delta T\frac{\mathrm{d}x}{\mathrm{d}s}\right), \tag{1}$$

$$\frac{\mathrm{d}^2 v}{\mathrm{d}t^2} = \frac{1}{m}\frac{d}{\mathrm{d}s}\left(T\frac{\mathrm{d}v}{\mathrm{d}s} + \Delta T\frac{\mathrm{d}y}{\mathrm{d}s}\right). \tag{2}$$

If the cable is inextensible—our experience with the statics of suspension bridge cables gives us good reason to suppose this to be a good approximate assumption—the geometry of the motion must conform to a further equation

$$\frac{dx}{ds}\frac{du}{ds} + \frac{dy}{ds}\frac{dv}{ds} = 0. \tag{3}$$

These three equations provide a basis for determining u, v and ΔT in terms of s and t, but they are intractable for a uniform cable with m constant, and so Routh adopted a condition more amenable to solution. He assumed that m so varied with s that the cable hung in the form of a cycloid. For the lowest natural frequency of such a cable, Routh then deduced

$$\lambda\alpha_0 = \pi, \tag{4}$$

where $\lambda = 2\sqrt{1 + (b\omega^2/g)}$ and $\sin \alpha_0 = l/8b$. Here b is the radius of the generating circle of the cycloid, l is the length of the chain, and ω is the circular frequency of the oscillation. For the case when α_0, the slope of the cable at its ends, is small, (4) can be written

$$n_1 = \frac{1}{2\pi} \sqrt{\frac{g}{b}\left(\frac{16\pi b^2}{l^2} - 1\right)}, \tag{5}$$

where n_1 is the lowest natural frequency concerned. Now if the dip to length ratio (d/l) is small, we can write

$$b = \frac{l^2}{32d}. \tag{6}$$

Substituting this in (5), and expanding the surd, we have

$$n_1 = \frac{1}{2\sqrt{2}} \sqrt{\frac{g}{d}}\left\{1 - \frac{32}{\pi^2}\frac{d^2}{l^2}\right\}, \tag{7}$$

where terms involving powers of d/l greater than the second are neglected. The higher frequencies given by Routh can be written similarly

$$n_2 = \frac{1\cdot43}{2\sqrt{2}} \sqrt{\frac{g}{d}}\left\{1 - \frac{32}{(1\cdot43\pi)^2}\frac{d^2}{l^2}\right\}, \tag{8}$$

$$n_3 = \frac{2}{2\sqrt{2}} \sqrt{\frac{g}{d}}\left\{1 - \frac{32}{(2\pi)^2}\frac{d^2}{l^2}\right\}, \tag{9}$$

$$n_4 = \frac{2\cdot5}{2\sqrt{2}} \sqrt{\frac{g}{d}}\left\{1 - \frac{32}{(2\cdot5\pi)^2}\frac{d^2}{l^2}\right\}. \tag{10}$$

All these results, of course, apply strictly only to a flat suspension cable of cycloidal form.

11.3. In 1948, the author, as a result of some simple experiments on cables, realised that n tended to vary inversely as $\sqrt{d}$ and was almost independent of the cable length l, and sought a simple reason for this. Stated in terms of dimensional theory, we may write

$$n = \phi\,(l, d, m, g), \tag{11}$$

and thence, by the usual dimensional arguments, come to the form

$$n = \left(\frac{g}{d}\right)^{\frac{1}{2}} \psi\left(\frac{d}{l}\right). \tag{12}$$

This conformed to available experimental findings provided the function ψ was such that the ratio d/l was almost ineffectual.

The simple approach[51] to the problem is to regard the oscillations of the cable as arising from the propagation of transverse waves (in a vertical plane) along its length. If the velocity of propagation v is treated as constant, the natural frequencies will be given by

$$n = i\frac{v}{l}, \tag{13}$$

where i has the values 1, 1·5, 2, . . . for the successive natural modes. Now for small amplitude waves the velocity v in a cable under tension T is given by

$$v = \sqrt{\frac{T}{m}}, \tag{14}$$

and since T does not vary greatly along a catenary with a small depth to span ratio, we expect (13) to give good results.

From (7) of Chapter 2, the tension at any point in a catenary is

$$T = mg \cdot y, \tag{15}$$

where y is measured from the directrix. At the lowest point of the cable $y = c$, the parameter of the catenary; to a close approximation, therefore, we can write for the average value T of the cable tension

$$T = mg(c + \tfrac{1}{3}d), \tag{16}$$

where d is the dip of the cable. Now the parameter of a catenary is given by

$$c = \frac{l^2}{8d} - \tfrac{1}{2}d, \tag{17}$$

and inserting this in (16) gives

$$T = \frac{mgl^2}{8d}\sqrt{1 - \frac{4}{3}\frac{d^2}{l^2}}. \tag{18}$$

By substituting from (18) in (14) and thence in (13), we have for the natural frequencies

$$n = \frac{i}{2\sqrt{2}} \sqrt{\frac{g}{d}} \left(1 - \frac{2}{3} \frac{d^2}{l^2} \right), \tag{19}$$

neglecting higher powers of d/l than the second. This is of the same form as equations (7) to (10) for the flat cycloidal cable, and, since d/l is in practice of the order of $1/10$, confirms the unimportance of the function $\psi(d/l)$ of (12).

As a result of an examination of the above results in the light of experiments, the author put forward, in 1949, the following semi-empirical formulae

$$n_1 = \frac{1}{2\sqrt{2}} \sqrt{\frac{g}{d}} \left(1 - 3 \frac{d^2}{l^2} \right), \tag{20}$$

$$n_2 = \frac{1 \cdot 4}{2\sqrt{2}} \sqrt{\frac{g}{d}} \left(1 - 1 \cdot 5 \frac{d^2}{l^2} \right), \tag{21}$$

$$n_3 = \frac{2}{2\sqrt{2}} \sqrt{\frac{g}{d}} \left(1 - 0 \cdot 7 \frac{d^2}{l^2} \right), \tag{22}$$

and demonstrated their accuracy by comparison with experiment.

It is of interest to note, in relation to the effects of moving vehicles on a suspension bridge, that the result (20) corresponds, from (13), to a wave velocity of

$$v = \frac{l}{2\sqrt{2}} \cdot \sqrt{\frac{g}{d}} \cdot \left(1 - 3\frac{d^2}{l^2} \right). \tag{23}$$

This gives, for $g = 32.2$ ft/sec² and a cable with $l/d = 10$,

$$v = 6.15\sqrt{l} \tag{24}$$

in ft/sec. Thus for a cable of span $L = 3000$ ft, corresponding, from (21) of Chapter 2, to $l = 3080$ ft, the travelling wave would have a velocity of 340 ft/sec.

11.4. The foregoing work excited the interest of Professor Satterly, of Toronto, and through him, that of Saxton and Cahn in U.S.A.,[52] who made a new mathematical attack on the classic problem. Taking advantage of the physics of the problem as set out above, and in particular of the importance of the transverse wave aspect of the motion, they reduced Routh's equations of motion (1), (2) and (3), for the case of a uniform cable, to the single equation

$$\cos \alpha \cdot \frac{d^4 \eta}{dt^4} - 2 \sin \alpha \cdot \frac{d^3 \eta}{dt^3} + \left(\cos \alpha + \frac{\lambda^2}{\cos^2 \alpha} \right) \frac{d^2 \eta}{dt^2}$$

$$+ \left(\frac{\lambda^2}{\cos^3 \alpha} - 1 \right) \cdot 2 \sin \alpha \cdot \frac{d\eta}{dt} - \frac{\lambda^2}{\cos^2 \alpha} \cdot \eta = 0. \qquad (25)$$

Here η is the displacement of the cable normal to its equilibrium configuration, α is its slope, and the boundary conditions are $\eta = d\eta/dt = 0$ when $\alpha = \pm \alpha_0$ at the cable ends; λ is given by

$$\lambda^2 = \frac{l}{2g \tan \alpha_0} \cdot \omega^2. \qquad (26)$$

The solution of (25) is then developed to lead to the explicit series results:

for odd modes

$$\lambda = \frac{i\pi}{A} \left\{ 1 - \frac{A(B + C)}{(i\pi)^2} + \ldots \right\}, \qquad (27)$$

and for even modes

$$\lambda = \frac{(i + \frac{1}{2})\pi}{A} \left\{ 1 - \frac{A(B + D)}{[(i + \frac{1}{2})\pi]^2} + \ldots \right\}, \qquad (28)$$

where $i = 1, 2, 3, \ldots$ and A, B, C and D are trigonometrical functions of α_0. When α_0 is small, (25) and (26) reduce to

$$\lambda = \frac{i\pi}{\alpha_0} \qquad (29)$$

and

$$\lambda = \frac{(i + \frac{1}{2})\pi}{\alpha_0} \left\{ 1 - \frac{1}{[(i + \frac{1}{2})\pi]^2} \right\} \qquad (30)$$

respectively.

It is of interest to study these results in relation to those of the elementary transverse wave theory. Let us compare, for example, (29) with (19) and (20). For $i = 1$, and α_0 small, (26) gives

$$\lambda_1 = \omega_1 \sqrt{\frac{l}{2g\alpha_0}}, \qquad (31)$$

and thence, from (29),

$$n_1 = \frac{\omega_1}{2\pi} = \frac{1}{\sqrt{2}} \sqrt{\frac{g}{\alpha_0 l}}. \qquad (32)$$

But α_0 is given approximately by $4d/l$, so that (30) can be written

$$n_1 = \frac{1}{2\sqrt{2}} \sqrt{\frac{g}{d}}, \tag{33}$$

which agrees with (20) if we neglect the small correction due to d/l.

In their paper, Saxon and Cahn compare the results given by the full expressions (27) and (28) with experimental results obtained by themselves, by Rudrick, Leonard and Saxon, and by the author; and, as will be seen in Fig. 28, excellent agreement is obtained, even for values of α_0 as large as 70°.

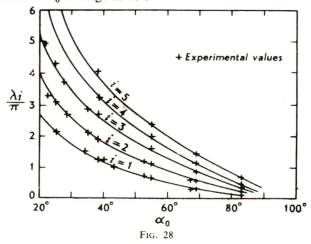

Fig. 28

11.5. Complete Bridge in Flexure

We have so far been concerned with the behaviour of the cable alone. Let us now consider such a cable stiffened by a girder as in a suspension bridge, and again restrict our attention to small oscillations in the vertical plane.

If we return to the linearised equilibrium equation for this structure, as given in (7) of para. 8.3, we have

$$EI \frac{d^4v}{dx^4} - H \frac{d^2v}{dx^2} = p - w \frac{h}{H}. \tag{34}$$

Now this equation will apply to the oscillation if we replace the external loading p by the reversed inertia loading $-(w/g)\,(d^2v/dt^2)$ and treat h as caused by the inertia forces and therefore a function of t. Providing h is small compared with H, this adaptation of (34) is justifiable, and we have as our linear equation of motion

$$\frac{w}{g} \frac{d^2v}{dt^2} + EI \frac{d^4v}{dx^4} - H \frac{d^2v}{dx^2} + \frac{w}{H} h = 0. \tag{35}$$

Writing $v = v_0 e^{\lambda t}$ and $h = h_0 e^{\lambda t}$, (33) reduces to

$$\frac{w}{g} \lambda^2 v_0 + EI \frac{d^4 v_0}{dx^4} - H \frac{d^2 v_0}{dx^2} + w \frac{h_0}{H} = 0. \tag{36}$$

Here v_0 is a function of x, and h_0 is connected to v_0 by an elastic relation. This equation is thus a linear one in v_0 of the fourth order with constant coefficients.

Consider first the case of antisymmetrical modes of oscillation, i.e. those involving an even number of half wave lengths and thus a node at mid-span. In this case, for small amplitudes, the motion involves a pendulum-like swing of the cable in its own plane and no additional cable tension h arises. As a result, the solution of (36) allowing for the end conditions appropriate to a simple span, becomes

$$v_0 = a \sin \frac{m \pi x}{L}, \tag{37}$$

where $m = 2, 4, 6, \ldots$ and the circular frequencies of the vibrations in these nodes are given by

$$\omega = \frac{m \pi}{L} \sqrt{\frac{g}{w} \left(H + m^2 \frac{EI}{L^2} \right)}. \tag{38}$$

To compare this result with those already obtained for the cable alone, we can substitute for H in terms of w and obtain for the frequency

$$n = \frac{\omega}{2\pi} = \frac{m}{2} \cdot \frac{1}{2\sqrt{2}} \sqrt{\frac{g}{d}} \cdot \sqrt{1 + 8 m^2 \pi^2 \cdot \frac{EId}{wL^4}}. \tag{39}$$

This is, in the form given by the author in 1949,[28] the result obtained by Steinman in 1943.[29] The term EId/wL^4 is our stiffness parameter R, so that we can write

$$n = \frac{m}{2} \frac{1}{2\sqrt{2}} \sqrt{\frac{g}{d}} \cdot \sqrt{1 + 8 m^2 \pi^2 R}. \tag{40}$$

When R is negligible, (40) degenerates to a result for the cable alone; for the lowest mode ($m = 2$), to

$$n_1 = \frac{1}{2\sqrt{2}} \sqrt{\frac{g}{d}}, \tag{41}$$

as in (33).

If we regard the quantity in the last part of (40) as a correction upon the cable frequencies for the presence of the stiffening girder, then, as in the author's paper of 1949, we can write

$$n_1 = \frac{1}{2\sqrt{2}} \sqrt{\frac{g}{d}} \left(1 - 3\frac{d^2}{l^2}\right) \cdot \sqrt{1 + 32\pi^2 R}, \tag{42}$$

$$n_3 = \frac{2}{2\sqrt{2}} \sqrt{\frac{g}{d}} \left(1 - 0 \cdot 7\frac{d^2}{l^2}\right) \cdot \sqrt{1 + 128\pi^2 R}. \tag{43}$$

A comparable result for n_2 is omitted from these results because it involves an odd value of m ($m = 3$) and therefore some change of h_0 in (34). This case has been studied at some length by Steinman[29] and by Bleich.[41] The former, concentrating on an inextensible cable, adopts a form in place of (37) of

$$v_0 = a \sin\frac{m\pi x}{L} - \frac{a}{m} \sin\frac{\pi x}{L}. \tag{44}$$

His solution on this basis alters the last term of (40) so that it reads

$$\sqrt{\frac{m^4 + 1}{m^4 + m^2} + \frac{m^6 + 1}{m^6 + m^2}} \cdot 8m^2\pi^2 R. \tag{45}$$

For the lowest symmetric mode $m = 3$; (45) then becomes

$$\sqrt{\frac{82}{90} + \frac{730}{738}} \cdot 72\pi^2 R.$$

Inserting this in (40) and rearranging, we have

$$n_2 = \frac{1 \cdot 43}{2\sqrt{2}} \sqrt{\frac{g}{d}} \cdot \sqrt{1 + 78 \cdot 3\pi^2 R}. \tag{46}$$

For the case when $R = 0$, the coefficient $1 \cdot 43$ here agrees very well with the semi-empirical one adopted by the author in 1949.*

Bleich's work on this problem, using the full equation (36), allowed for the elastic extensibility of the cable and led to much more complicated expressions that cannot be solved explicitly. His energy solution for the lowest symmetrical mode ($m = 3$) assumed the form

$$v_0 = a_1 \sin\frac{\pi x}{L} + a_3 \sin\frac{3\pi x}{L}. \tag{47}$$

On this basis,

$$h_0 = \frac{16d}{\pi L} \frac{E_c A_c}{l} \left(a_1 + \frac{a_3}{3}\right), \tag{48}$$

and the frequency equation reduces to

$$\frac{32d}{\pi^3} A + 9 \left(\frac{32d}{\pi^3} - \frac{a_1 + \dfrac{a_3}{3}}{h} \cdot A\right) B = 0, \tag{49}$$

---

* See equation (21), p. 112.

where

$$A = \frac{wL^2}{g\pi^2}\,\omega^2 - H - \frac{\pi^2EI}{L^2}, \tag{50}$$

$$B = \frac{wL^2}{g\pi^2}\cdot\omega^2 - 9H - 81\,\frac{\pi^2EI}{L^2}. \tag{51}$$

In the case when the cable is inextensible, h becomes, from (48), very large, and (49) reduces to

$$\alpha^2 = \frac{8{\cdot}2H + 73\,\dfrac{\pi^2EI}{L^2}}{\dfrac{wL^2}{g\pi^2}},$$

which can be written as

$$n_2 = \frac{1{\cdot}45}{2\sqrt{2}}\sqrt{\frac{g}{d}}.\,\sqrt{1 + 69{\cdot}3\pi^2R}, \tag{52}$$

which approximates to Steinman's earlier result.

From a study of equation (49), Bleich shows that in practice the extensibility of the cable can have an appreciable effect on the frequency n_2 and should be allowed for, particularly when R is large (i.e. girder relatively stiff).

Throughout this and the foregoing sections of this chapter, we have based our natural frequency calculations upon the fundamental differential equations. This is necessary for the general treatment attempted in this chapter, but for any particular bridge we could have proceeded by considering the whole structure, cable and stiffening girder, in terms of a number of discrete masses on appropriate springs, represented by a flexibility or stiffness matrix. Such a treatment lends itself to rapid numerical solution by digital computer, as has been shown by N. K. Chaudbury and D. M. Brotton.[53]

11.6. Complete Bridge in Torsion

If we turn now to torsional oscillations, we have first to note that in many suspension bridges the lateral deck structure (wind bracing) is in one plane only and so the torsional stiffness of the deck itself is small. In such cases the natural modes of the bridge in torsion correspond precisely to the natural modes of vertical vibration just discussed, except that the two sides of the deck will be moving in opposition, i.e. 180° out of phase. The lowest, and most important, torsional mode thus becomes one having nodes at the ends and at the centre of the span only, being the torsional counterpart of the flexural mode n_1 of equation (42).

Certain differences between the frequencies of these two comparable modes—flexural and torsional—arise, however, from the different inertial conditions. In the flexural mode, the vertical motion of the deck is uniform across any one cross-section; in the torsional mode, one side is rising when the other is going down and the mid-point of the deck is stationary. This brings the radius of gyration of the deck section into the analysis for the torsional case.

There is another difference which we propose to neglect here. In the mode concerned, each cable (or group of cables on one side of the bridge) is swinging in pendulum fashion in its own plane. This means that when one cable has a horizontal movement in one direction along the span, the other on the other side of the bridge has an opposite movement. As a result, the deck will, apart from restraints at its ends, tend to *yaw* in its own plane, i.e. oscillate about a vertical axis through its centroid. However, this effect will be small so long as the vertical suspension rods, particularly those in the middle, are not short.

To allow for the torsional rotation of the deck mass, we have first to distinguish between the weight of the cables and the weight of the deck. Let

w_D = weight per unit length of deck per cable (= half total weight
      of deck per unit length, assuming a two cable system),
w_c = weight per unit length of one cable,

the length in both cases being measured along the span. Then if k_D is the radius of gyration of the deck section, in the flexural case the weight involved in the vertical motion at one side of the bridge is

$$w_D + w_c,$$

and in the torsional case it is

$$\bar{w} = \frac{4k_D{}^2}{b^2} w_D + w_c, \tag{53}$$

where b is the breadth of the deck between the cables. As the frequencies are inversely proportional to the square roots of these weights, the torsional frequency n_T will be related to the corresponding flexural frequency n_F by the relation

$$\frac{n_T}{n_F} = \sqrt{\frac{w_D + w_c}{\bar{w}}}. \tag{54}$$

The radius of gyration k_D is commonly of the order of $b/3$, so that if, for example, w_D is twice w_c, n_T is some 25 per cent greater than n_F. Equation (54) thus provides a ready means of estimating the torsional frequencies related to each of the flexural frequencies already discussed, provided always that the torsional stiffness of the deck structure is small.

11.7. When the deck structure is relatively stiff in torsion, as occurs when the wind bracing is at two levels, above and below the roadway, or when the deck is itself a torsion box in the form of a tubular framework or a thin-walled tube of plating, this stiffness can have substantial effects on the torsional frequencies of the bridge.

If we restrict attention to the lowest relevant frequency—that involving nodes at the ends and at the centre only—we can say, as in paragraph 11.5, that no increase in the cable tension is involved, and can write for the mode concerned

$$\zeta = a \sin \frac{2\pi x}{L}, \tag{55}$$

where θ is the angle of twist along the deck at any one time.

Our problem in this form has been solved by Bleich[41] for a bridge with a uniform tubular framework and by Selberg[54] for one with a uniform thin tube.

In both cases the analysis is somewhat lengthy and difficult to cast in a general form of clear physical significance. We therefore adopt here an approximate analysis given by Y. Rocard[55] that brings in both the flexural and torsional stiffnesses of any tubular deck without discussing the detailed structural actions, such as warping of cross-sections, that must occur in practice.

For the unrestrained torsion structure, we may write for the twist θ due to a given uniform torque T

$$T = GK \cdot \frac{d\theta}{dx}, \tag{56}$$

where K is the effective polar second moment of area for the structure.

In the case of a thin-walled tube, K is given by

$$K = \frac{4A^2}{\int \frac{ds}{t}}, \tag{57}$$

where A is the area enclosed by the walls, t is the wall thickness, and s is measured around the walls of the tube.

Here GK represents the torsional stiffness of the whole deck structure, to which Rocard adds a flexural stiffness EI (as of a girder) below each cable. On this basis, and assuming that the deck is suspended by vertical hangers offering no resistance to relative longitudinal displacement between cables and deck, the differential equation for torsional motion of the bridge becomes

$$4\,\frac{\bar{w}}{g}\cdot\frac{k_D{}^2}{b^2}\cdot\frac{\mathrm{d}^2\theta}{\mathrm{d}t^2}+\frac{1}{2}\,EIb\cdot\frac{\mathrm{d}^4\theta}{\mathrm{d}x^4}$$

$$-\left(\frac{\bar{w}\,.\,L^2b}{16d}+\frac{GK}{b}\right)\frac{\mathrm{d}^2\theta}{\mathrm{d}x^2}+\frac{8d}{L^2}\,.\,h=0. \qquad (58)$$

This equation is the torsional equivalent of the flexural equation (35), and its solution, for the mode represented by (55), gives for the circular frequency in torsion

$$w_T{}^2=\frac{b^2g}{4k_D{}^2}\left\{\frac{16\pi^4EI}{\bar{w}L^4}+\frac{\pi^2}{2d}\right\}+\frac{2\pi^2\,.\,GK\,.\,g}{\bar{w}\,.\,k_D{}^2L^2}. \qquad (59)$$

Here the middle term represents the effects of the cables; when $EI = GK = 0$, equation (59) breaks down to the result (33) for the cables alone. The first term expresses the influence of the flexural stiffness and the last that of the torsional stiffness of the deck. The whole thus gives a means, for a particular bridge design, for assessing the contributions of the various components to the torsional frequency. Rocard points out the special importance of the last term in bridges of very long span.

CHAPTER 12

Oscillations of Suspension Bridges
under Lateral Winds

12.1. [Until recent decades, whilst the tendency of suspension bridges to oscillate under high lateral winds was known to designers, little provision for wind action was made beyond that necessary to withstand the static lateral forces that may arise. [With the spectacular collapse of the Tacoma Bridge in the U.S.A. in 1940,[25] however, the aerodynamic resources and research methods of aeronautics have been brought to bear on the subject and have produced a succession of valuable contributions to its understanding.] Of these the most extensive are the model investigations of F. B. Farquharson[56] in the U.S.A. and of R. A. Frazer and C. Scruton[26] in England, and the theoretical work of F. Bleich.[27] A more recent publication[41] of the U.S. Department of Commerce gives an excellent review of the mathematical theory of suspension bridge vibrations. D. B. Steinman,[29] in 1943, published a valuable summary of relevant vibration data from the practical design standpoint. The present state of knowledge in this whole field was well covered at a recent symposium[57] at the National Physical Laboratory, when reference was made to relevant work in Norway (by A. Selberg) and France (by Y. Rocard).

12.2. Static Actions

[Let us first consider in brief the static effects of a steady lateral wind.] An isolated uniform suspension cable would be subject, under such conditions, to the combined action of uniformly distributed gravity loading in a vertical plane and uniformly distributed wind loading acting horizontally. The overall effect would be to swing the whole cable so that it hung not in a vertical plane but in a plane inclined to the vertical in such a way as to contain the resultant force on every element of the cable length.

[In the case of complete suspension bridges, however, the greatest wind forces commonly arise on the bridge deck, which tends to resist them, on its own account, as a beam bending in a horizontal plane.] The wind bracing in the plane of the deck is provided to resist the shear forces thus arising and the stiffening girders at each side of the deck act in part as the booms or flange members of this beam.] To enable the deck structure to resist wind forces in this fashion, the deck is usually specially supported against horizontal movement at each end (at each tower in the case of a multi-span bridge). [The deck beam, however, is usually very flexible against lateral wind forces, and its lateral displacement at the span centre is large enough to involve, due to the presence of the inextensible cables above, a lifting of the deck at its centre.] Thus in a complete bridge the wind forces are resisted partly by the elastic flexure of the deck structure in a horizontal plane and partly by gravity action induced by the cables.

An early analysis of this problem was given by D. B. Steinman.[58] For the case of a single span, if we concentrate attention on conditions in the region of the centre, the lateral deflection of the deck there due to a net lateral loading p' will be

$$\delta = \frac{5}{384} \frac{p'L^4}{EI} = 0.013 \frac{p'L^4}{EI},\tag{1}$$

where EI refers to the flexural stiffness of the whole deck for bending in the lateral plane.

Here p' will be less than the total wind loading p per unit span* because of the gravity resistance resulting from the swing δ of the total gravity loading w acting approximately at deck level, a depth D below the tower tops. On this account, the gravity resistance $p - p'$ will be equal to $w . (\delta/D)$. Hence, from (1), we have

$$\frac{p - p'}{p} = \frac{r}{p} = \frac{0.013 \dfrac{wL^4}{DEI}}{1 + 0.013 \dfrac{wL^4}{DEI}}.\tag{2}$$

Steinman points out that p' will approximate to p at the ends of the span and so recommends for the design of the deck an average effective horizontal loading at deck level of

$$p'' = p - \tfrac{5}{6}r.\tag{3}$$

In his book[11] he gives examples showing that p'' may be some 10 or 20 per cent less than p.

Subsequently, Moisseiff and Lienhard[59] studied this lateral problem in more detail, though agreeing with the value of the above approximate treatment. More recently, Arne Selberg[60] has investigated the same problem more deeply. By expressing the gravity resistance $p - p'$, or $p - p''$, in terms of a Fourier series, he shows how a variety of suspension bridge arrangements can be solved for the static wind loading case.

---

* We here neglect that part of the wind loading on the cables that is carried directly to the tower tops.

Before proceeding to the consideration of oscillations as such, it should be noted that real winds, because of their eddying and gusty nature, do not produce "static actions." An extensive study of the effects of such fluctuations has been made by A. G. Davenport,* first at Bristol and later in Canada. He has shown that on long span bridges turbulent winds can produce stresses in the deck structure of significance in design from both strength and fatigue aspects.

12.3. General Nature of Oscillations

[We are more concerned here, however, with the oscillation problems of suspension bridges. That such bridges sometimes suffer serious damage due to oscillations in high winds has been a matter of common knowledge from early times, and technical records of bridge failures, starting with that of a footbridge of 260 ft span over the Tweed at Dryburgh in 1817, are numerous in the nineteenth century.] Nevertheless, it was not until such troubles reached a climax in 1940 with the dramatic collapse of the 2800 ft span bridge at Tacoma Narrows in a wind of only some 40 miles per hour that much serious scientific attention was devoted to the problem.

It is first necessary to distinguish between three types of oscillatory motion:

(a) Purely flexural oscillations of the bridge, with each cross-section of the bridge deck moving up and down in a vertical plane, every point in it having the same amplitude of motion.

(b) Purely torsional oscillations of the bridge, with each cross-section of the bridge deck oscillating angularly in its own plane about an axis at or close to the mid-line along the roadway.

(c) Coupled flexural-torsional oscillations of the bridge, with each cross-section of the bridge deck undergoing in general both vertical and angular motions.

In gusty weather a lateral swinging of the whole bridge deck about the tower tops will sometimes arise, but in modern long span bridges this form of oscillation is of small amplitude and short-lived.

All three types of oscillation listed above are liable to occur only under a lateral wind having its main component at right angles to the vertical plane of symmetry of the bridge, and the effects of such a wind are usually somewhat aggravated if the wind, instead of acting horizontally, has a slight upward inclination. Oscillations of type (a) are common among suspension bridges built in the nineteenth century, but are usually of harmless amplitude. Oscillations of type (c) are rare, but may be dangerous. Oscillations of type (b) are not so common as (a), but when present may build up

---

* See, for example, Ref. 61.

to dangerous amplitudes and precipitate coupled oscillations of type (*c*).

In order to understand the nature and origin of these various oscillations, it is desirable first to consider the air forces on a bridge deck under a lateral wind. In the simplest possible case we may

regard a deck as simply a flat plate or ribbon of metal running along the bridge span, and consider the action of a lateral airstream upon a section of its length. Provided the wind is in a nearly horizontal plane, the bridge deck section will in such a case act like a simple thin aerofoil. The flow across the deck will be of a smooth stream line character, as in Fig. 29, and the deck will experience a lateral *drag* force together with a small vertical *lift* force, both varying linearly with the inclination of the wind to the plane of the deck. These forces will be steady for a given wind speed and any small flexural or torsional oscillation given to the deck will normally be found to die out due to positive damping arising from the air forces as modified by the motion and from the natural positive damping

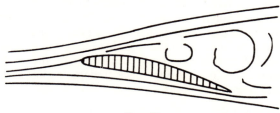

in the bridge structure, due mainly to friction. The only possible sources of unstable oscillations in this case would be the effects of a coupled oscillation of type (*c*) or due to the development of unsteady flow, with possible negative aerodynamic damping, if the wind inclination to the horizontal became large enough for the streamline flow of Fig. 29 to break down and *stall* as in Fig. 30. In such a case large eddies would be shed somewhat irregularly from the aerofoil or deck and cause the air forces on the deck to fluctuate; for the simple flat plate deck, however, the necessary wind inclination or "incidence" would be about 10° and unlikely to occur for any length of time in nature.

Much more important in practice are the effects of the stiffening girders at the sides of the deck. If these are plate girders—as was common in the nineteenth century—with continuous webs, they present a uniform bluff obstacle to any lateral wind and induce

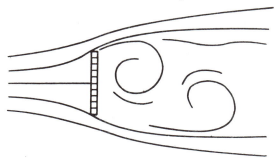

FIG. 31

eddying flow at all speeds. This is very clear from the flow picture of Fig. 31, which illustrates the effects of a single long flat plate normal to the wind. In this case large eddies or *rollers* leave the top and bottom edges of the plate alternately, and so produce regular periodic pulsations in the air forces on the plate. Fig. 32

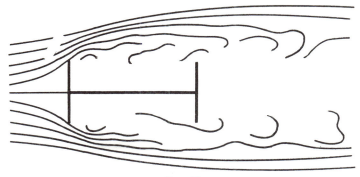

FIG. 32

illustrates the flow past a model suspension bridge deck with plate girders at its sides, and the possibility of fluctuating forces on the deck itself as well as on the side girders is at once evident. Now the eddy frequency in the idealised case of Fig. 31 is found to vary linearly with the wind speed V and inversely as the depth D of the obstacle, and is given approximately by the formula*

$$f = 0 \cdot 2 \frac{V}{D}. \tag{4}$$

* See, for example, Ref 62.

There is at once a possibility of resonance between this eddy frequency f and some natural frequency n, in either flexure or torsion, of the bridge itself. Moreover, from the eddying nature of the air flow it is possible that, as in the stalled case of Fig. 30, the aerodynamic damping, particularly of the torsional motion, will be negative and so offset any natural positive damping, due to friction, in the structure.

There are here the elements of an explanation, though oversimplified, of oscillations of types (a) and (b). At a critical wind speed V_c the frequencies f and n would, from (4), match* when

$$V_c = 5nD. \tag{5}$$

It is clear that in practice both the depth of girder D and the width of deck B must influence the eddy formation, its regularity and its effects; and since for many bridges the ratio B/D will not vary much, it has become customary to express the critical wind speed V_c in terms of B, thus,

$$V_c = knB, \tag{6}$$

and to examine full scale and model evidence in terms of a non-dimensional *reduced velocity*

$$V_r = \frac{V}{nB}. \tag{7}$$

This *reduced velocity* is a mode of presentation common in the *flutter* work of aeronautics, though not perhaps quite so directly relevant there, and has become usual in this bridge application through the work of Bleich and Frazer. Equation (6) also has a direct aeronautical parallel. In the early flutter investigations of the decade before the last war, it was common to use for the critical flutter speed an approximate formula like (6) attributed to Kussner,[63] and the same formula, in reduced frequency form, has been advocated recently for preliminary design work, by Broadbent.[64] The *constant k* is, of course, regarded as a parameter to be varied with other influential factors, such as, in the bridge case, the ratio B/D.

12.4. Flexural Oscillations

With the foregoing aerodynamic and aeronautical notions in mind, we can now examine in more detail oscillations of types (a) and (b). Let us concentrate first on (a), the purely flexural oscillations, and restrict our attention to a bridge deck with plate girders at its sides. As the lateral wind increases in speed, the magnitude and frequency of the fluctuating air forces on the deck will increase, and the bridge deck will gradually start to oscillate in flexure at its

---

* See Ref. 28 and its references.

lowest natural frequency (n_1 of Chapter 11). If the eddies shed by the windward girder are regular enough, the amplitude of the oscillations will tend to become greatest when resonance occurs, but in general the limit of amplitude at a given wind speed will be set by the natural aerodynamic and structural damping present. There is here no inherent instability in the oscillation—i.e. no tendency for the oscillation, at a given wind speed, to grow indefinitely in amplitude—and the effects of the oscillation are likely to be felt more as an inconvenience to traffic than as a serious danger to the structure. At higher wind speeds, the mode and frequency of oscillation may change to a higher natural frequency in flexure, such as n_2 or n_3, but in such cases the structural damping will tend to become more restrictive of the stable amplitude.

Experiments on small scale models of short lengths of bridge decks (*section models*), suitably sprung to give vertical motion, and on complete bridge models, show that the amplitude of these flexural oscillations is a function of D/B. It is only when D/B is large (over 0·2) that the amplitudes reached are likely to be catastrophic. This increase of danger with D/B is, of course, consistent with our rough notions based on the eddies shed by the windward girder.

It remains to consider how, if possible, we can prevent these flexural oscillations. One way, of course, is to minimise them by adopting deck designs with D/B small, but a much more effective way is to reduce the scale of the eddies not by reducing D but by abandoning a plate girder in favour of a lattice or other open web truss. The scale of the eddies in such a case is set by the width of the individual bracing members instead of by the overall depth of the girder, and model tests have shown that the flow over the bridge deck then becomes so irregular and broken up that either no flexural oscillations occur or their amplitude is much reduced. This reveals a principle to be adopted more generally: footpath handrail girders should be truss-like, road safety barriers should be open, and so on, all with the idea of breaking up the flow and minimising its effect on the deck as a whole.

Selberg[54] and others have found by experiment that these vertical oscillations tend to occur over two ranges of wind speed, with a stable range in between. In the terms of equation (6), it appears that the unstable regions commonly occur between the speeds

$$V_1 = 0.25\, nB \text{ and } V_2 = 0.45\, nB,$$

and again between

$$V_3 = 3.0\, nB \text{ and } V_4 = 4.0\, nB;$$

but in neither case, and particularly when open web trusses are used, are the oscillations produced likely to be of amplitudes more than unpleasant to pedestrians. Many existing suspension bridges exhibit over some limited range of wind speed moderate vertical oscillations of this character that are structurally harmless.

12.5. Torsional Oscillations

We can consider purely torsional oscillations of type (b) on the same lines. For solid decks with plate girders at the sides, however, torsional oscillations do not start and just grow with the wind velocity. The behaviour is more sensitive to wind speed and incidence, and to the ratio D/B, and oscillations tend to start at some critical wind speed V_c in the fundamental torsional mode n_T of Chapter 11. In this case, therefore, an expression such as (6) becomes directly relevant, with k dependent on D/B. Experiments on *section models* sprung to allow of torsion show that k increases considerably as D/B decreases. Thus for one series of such models, as D/B decreased from 0·29 to 0·025 (practically a flat plate), V_c rose from 12 miles per hour to 65 miles per hour, the oscillation frequency remaining at n_T throughout.

Now that it is known that with truss type stiffening girders flexural oscillations of type (a) are very unlikely, research effort on the aerodynamics of suspension bridges has largely been concentrated on the behaviour in torsion of decks with truss type girders. This is most conveniently done by model experiments in wind tunnels, using sectional models arranged so that the deck is broadside on to the air stream and given freedom to pitch about a longitudinal axis on the model centre line at a known frequency governed by suitable springs. It is found that the torsional or pitching stability of such models is not very sensitive to the level of the pitching axis relative to the roadway, but normally improves somewhat for axes above or below the deck level. In the absence, therefore, of precise knowledge of the natural axis of rotation, tests are best made with the axis on the road level so as to ensure *safe* results.

An interesting series of tests of this type on some deck arrangements first contemplated for the Severn Bridge and later applied to the design of the deck of the Forth Bridge have been described by Scruton,[65] and his paper indicates the qualitative effects of a variety of possible variations. Among these were the marked stabilising effects of longitudinal gaps in the roadway between traffic lanes, but other tests, such as those by Selberg,[54] have shown that such gaps do not act in this way for all deck arrangements.

The aim in the model work is in each case to determine an appropriate geometrical deck shape such that under lateral winds of all practicable speeds the deck section is stable for torsional oscillations about a longitudinal axis near the road level at its centre line.

In other words, if the deck is caused by any extraneous forces to start to oscillate in torsion at its natural frequency, such oscillations will be damped and die away instead of growing, whatever the wind speed.

12.6. Some consideration must be given to the scales adopted for such models. Much of the English work has been done on models of linear scale $\frac{1}{100}$ and $\frac{1}{32}$. The larger the scale, the easier it is to represent geometrically the details of construction and the less likely are errors due to aerodynamic scale effects. The latter effects are due to differences of flow pattern liable to arise between model and full scale, and are associated fundamentally with the viscosity of the air. Ideally, on this count, the Reynolds Number, which is VB/ν, where ν is the air viscosity concerned, should be the same for both model and full scale. This is commonly not practicable in aerodynamic experiments, but for bridge model scales in the above-mentioned region the errors arising are believed to be small.

It is not usual, in designing a bridge section model, to adopt values of the inertia and torsional spring stiffness to give true dynamic similarity with full scale. This is unnecessary because the frequency of oscillation of the model, as of a bridge, is found to be almost independent of the wind speed and so approximately constant at the natural frequency n_T of the model on its springs. Thus in planning section model tests the torsional stiffness (elastic) to be provided is partly one of convenience; in fact, the scales for speed and frequency are interdependent, but one or the other may be chosen at convenience. Expressing wind speed, therefore, in terms of the reduced velocity V_r, we may write in dimensional terms

$$V_r = \psi(\rho, \delta, I, B), \tag{8}$$

where ρ is the air density, B is the breadth of the deck, as before, I is its moment of inertia about the torsional axis, and δ represents the mechanical damping of the torsional motion and is its logarithmic decrement ($=$ natural logarithm of the ratio of successive amplitudes of torsional oscillation in still air). The terms in the function ψ in (8) may be grouped into the non-dimensional parameter $I\delta/\rho B^4$ so that, since ρ will usually be the same for model and full scale conditions, V_r will, for geometrically similar deck shapes, depend only on $I\delta/B^4$. Thus for equality of V_r in model and full scale conditions, using dashed symbols for the model

$$\frac{I'\delta'}{B'^4} = \frac{I\delta}{B^4}. \tag{9}$$

It is found not difficult to achieve this relationship approximately, and a precise statement of critical reduced velocity results can refer to the exact value of full scale damping (δ) to which they apply.

The decrement δ for actual suspension bridges is in the range 0.02 to 0.20, and a number of full scale measurements are now available.[66] High values are of special value for the reduction or suppression of simple flexural or torsional oscillations. The wooden decks of the early light suspension bridges gave δ values above 0.10; in some modern bridges structural dampening has been increased by the adoption of inclined rather than vertical hangers.

12.7. Coupled Oscillations

It remains to give some consideration to the possibility of coupled oscillations of type (c) in para. 12.3. These are oscillations of the type referred to as *flutter* in the classical aeroelasticity work of aeronautics. Flutter of this kind arises essentially from the coupling, or interdependence, of two or more degrees of freedom and in the case of a bridge deck would involve motion of both the vertical or flexural type (a) and the torsional type (b). It is usual—and indeed normally mathematically only practicable—to discuss such coupled oscillations in terms of linear equations for small amplitude movements, and we will follow this procedure in the present case. On this basis, the purely vertical motion of our bridge deck, expressed in terms of one mode of motion specified by the displacement v at a particular section, will be given by the equation

$$A_1\ddot{v} + B_1\dot{v} + C_1v = 0, \tag{10}$$

where $\ddot{v}$ and $\dot{v}$ are the second and first differentials of v with respect to time, A_1 is an effective moment of inertia, B_1 a damping term which, in the absence of mechanical damping, may be taken to be proportional to the wind speed V, and C_1 is the effective elastic stiffness of the bridge in flexure. Equation (10) is thus the linearised expression of our oscillation type (a) of para. 12.4. If a small disturbance in flexure is given to the bridge, (10) describes the resulting oscillation in flexure, assuming that alone is involved, and this oscillation will decay or grow according as B_1 is positive or negative.

Similarly, we may describe our torsional oscillation (b) in this linearised form in terms of the twist θ by the equation

$$A_2\ddot{\theta} + B_2\dot{\theta} + C_2\theta = 0, \tag{11}$$

and the stability of this at a given wind speed will again depend on the sign of the damping term B_2. In general, however, motion of one kind will affect motion of the other, and this interaction is expressed in classical aeroelasticity by the addition of further terms to equations (10) and (11), thus

$$A_1\ddot{v} + B_1\dot{v} + C_1v + D_1\ddot{\theta} + E_1\dot{\theta} + F_1\theta = 0, \tag{12}$$

$$A_2\ddot{\theta} + B_2\dot{\theta} + C_2\theta + D_2\ddot{v} + E_2\dot{v} + F_2v = 0. \tag{13}$$

Equations (12) and (13) are in this case simultaneous differential equations governing the resultant motion, involving both v and θ. The coefficients D_1, E_1, F_1 and D_2, E_2, F_2 represent *couplings* between the pure motions v and θ. The inertial couplings D_1 and D_2 are of the nature of products of inertia and, for a symmetrical bridge, will be small or zero, and the most important coupling term is likely to be F_1. In the case of aeroplane wings, both the products of inertia and F_1 are particularly important in practice.

There are a number of standard ways of discussing the stability of the coupled motion represented by equations (12) and (13). The earliest employs Routh's discriminants,[50] and this approach has been generalised by the test functions of Frazer.[67] But in any particular and not too complex a case, the motion itself can be studied in a physical manner and the nature of its stability examined.* Thus in the present case, where a number of the coefficients of (12) and (13) are functions of V or V^2, we can consider the possibility of instability (growing oscillations) by examining the critical speed V, if any, at which a state of neutral equilibrium (oscillations of fixed amplitude) is possible. In such a state, we may write

$$v = v_0 \sin(\omega t + \varepsilon), \tag{14}$$

$$\theta = \theta_0 \sin \omega t, \tag{15}$$

where v_0 and θ_0 are the amplitudes of the initial disturbance, ω the circular frequency of the steady oscillation, and ε the phase difference between the two types of motion involved. We have here assumed, as is reasonable, that the motion will be sinusoidal. By inserting (14) and (15) in (12) and (13), we can rewrite these equations of motion in terms of $\sin \omega t$ and $\cos \omega t$, and thence argue that, for neutral equilibrium, the coefficients of $\sin \omega t$ and $\cos \omega t$ in each transformed equation (12) and (13) must be separately zero if the equations are to hold for all values of t. Thus four simultaneous equations are derived for the four unknowns ω, ε, amplitude ratio θ_0/v_0, and the speed V at which this condition of neutral equilibrium arises.

It is found, as a result of the above process, that the frequency of the coupled motion will lie between the two natural frequencies n_F and n_T of the system, as given by equations (10) and (11) respectively when $V = 0$; one of the main arguments for believing that this coupled type (*c*) oscillation may occur in suspension bridges is the fact that some models have revealed oscillations with

---

* See, for example, Ref. 68.

frequencies between n_F and n_T instead of closely equal to n_F or n_T (in any natural mode).

The linear theory, however, as so far outlined, which is suitable for treating the smooth flow over thin aerofoils, has obvious limitations in relation to the eddy motion over bridge decks; Bleich[27] has tried to allow for this, and at the same time provide a general theory to include all three types of oscillation, by adding to equations (12) and (13) sinusoidal forcing terms representing pulsating vertical forces at the windward stiffening girder. The theory so developed is the best at present available, but as our knowledge of the aerodynamic coefficients of (12) and (13) is still very inadequate, let alone the coefficients in the forcing terms, designers will for long be forced to turn to empirical methods for preliminary design and thereafter, in all important cases, to resort to sectional model and other oscillation tests to check the stability of the resulting bridge.

12.8. One of the best developed empirical treatments of flutter in suspension bridges is that put forward by Selberg.[54,69] He takes as a standard of reference the flutter speed V_F for a simple "flat plate" deck, and then assesses the critical flutter speed V_{crit} for any actual design of deck by writing

$$V_{crit} = k \cdot V_F, \qquad (16)$$

where k is based upon available wind tunnel tests and is usually less than unity. For decks with open truss stiffening girders, k is most affected by the incidence of the lateral wind and by the magnitude of the initial disturbance. Fig. 33 illustrates the variation of k with these parameters for a particular bridge section.

As mentioned at the end of para. 12.3, equation (6) has sometimes been used for the direct approximate evaluation of the flutter speed of an aeroplane wing. Selberg's empirical expression for V_F is an elaboration of the same equation, and in the terminology of this book can be written

$$V_F = 3.9 \, n_T B \, \sqrt{\left\{1 - \left(\frac{n_F}{n_T}\right)^2\right\} \frac{\nu^{\frac{1}{2}}}{\mu}}, \qquad (17)$$

where n_F and n_T are the natural frequencies of the bridge in flexure and torsion respectively, and ν and μ are non-dimensional mass parameters defined by

$$\nu = 8 \left(\frac{k_D}{B}\right)^2, \qquad (18)$$

$$\mu = \frac{\pi \rho B^2}{4 \bar{w}}. \qquad (19)$$

Here ρ is the density of the air, k_D as before is the radius of gyration of the deck section, and

$$\bar{w} = w_D + w_C \tag{20}$$

as in equation (54) of para. 11.6.

For practical design, it is desirable that V_F should be as high as possible. The dominating variables in (17) are n_T and B, and both of these should therefore be kept large. Thus the modern tendency towards decks with tubular structures providing for four or more lanes of traffic is an automatic safeguard against flutter.

The way the term $\dfrac{n_F}{n_T}$ appears in equation (17) illustrates the great desirability, well known in binary flutter theory, that the basic natural frequencies should be well apart. This is commonly easy to achieve in suspension bridges. Of less importance, but of interest in relation to an early belief that suspension bridges were too heavy for the wind to excite a flutter, is the indication through the parameter μ of the beneficial effect of the weight of the bridge.

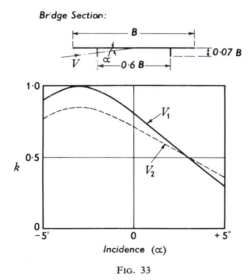

FIG. 33

V_1 = wind speed for just stable oscillation initiated by torsional oscillation of ± 0.01 rad.
V_2 = wind speed for just stable oscillation initiated by torsional oscillation of ± 0.1 rad.

CHAPTER 13

Towers

13.1. The towers of modern suspension bridges are broadly of three types:

(i) Stiff towers (as of masonry or cast iron), fixed at their bases, that support the main cables through a "carriage" that is free to slide or roll horizontally on the tower top, so that the tower itself is required to offer resistance only to the vertical components of the cable pulls.

(ii) Towers that are hinged at their bases so that they are free to rock in the plane of the main cables, which are securely attached to the tower tops. The towers themselves are thus simple struts (as of steel or reinforced concrete) between the cables and the tower foundations, and offer no resistance to spanwise movements of the cables.

(iii) Towers that are fixed at their bases and have the main cables secured to the tower tops. Such towers (of steel or concrete) are essentially vertical cantilevers, and offer some resistance (usually small) to cable movement spanwise.

In all three types, the towers are usually called upon to support not only the main cables at the tower tops but also the bridge deck girders at a lower level.

Many early suspension bridges, and some short span (up to say 500 feet) modern ones, have had towers of type (i). The long span bridges of this century have in most cases had towers of type (iii), but a few bridges of intermediate span, and one quite large bridge (Florianopolis, with a main span of 1114 feet) have had rocker towers of type (ii). Because of their special relevance to most long span suspension bridges, we shall concentrate here on fixed base towers of type (iii).

13.2. Towers of this kind are subject to a number of loading actions. The principal one is, of course, the vertical load applied at the tower top by the main cables. Added to this there will also be the vertical loads applied by the road girders at a lower level (the road supports are usually such that any horizontal forces in the road girders are resisted not at the towers but at the anchorages or the approach structures). Wind loads, both direct and indirect via cables and deck girders, have also to be treated.

All these loading actions are, superficially at least, common to many bridge structures. What is peculiar to suspension bridge towers is the way in which cable movements and cable extensions, as by temperature changes, act upon the towers. They force the tower top to move in a spanwise direction; and so bend the tower like a cantilever. This causes the large vertical load at the top to acquire a substantial eccentricity relative to the tower base. Thus, however straight and vertical a tower may be initially, the centroid of its section at the top will become offset spanwise from that of the bottom section; for bridges of 3000 feet span, such offsets may sometimes be 2 feet.

There are two possible approaches to the resulting design problem; firstly, to use a stiff tower of low or moderate slenderness ratio so as to resist the cable movement and keep down the net offset; or, secondly, to use as flexible a tower (high slenderness ratio) as one dare so as to reduce its resistance to movement and so minimise the net horizontal force applied by the cables.

There have been advocates of both policies; the earlier long span bridges in the U.S.A. had fairly stiff towers; the more recent ones there have tended to more flexible towers; and the towers of the Forth and Severn Bridges are still more flexible.

A limit to this process occurs when the stiffness of the tower is such that, were the tower top free, the tower would become unstable under the vertical load W applied by the cables. For towers of uniform section, this would occur when

$$W = \frac{\pi^2 EI}{4h^2}, \tag{1}$$

where EI is the flexural stiffness of the tower and h is its height. Under these conditions the tower would offer no horizontal resistance to the cables. If we can, at the same time, assume that the cables effectively prevent any further horizontal movement beyond that natural to them (as though supported on rollers on a rigid tower), the tower itself will, so far as the load W is concerned, be like a column fixed at its root and held by a hinge at its top. The critical Euler load for such a column is just under

$$P_E = \frac{2\pi^2 EI}{h^2}. \tag{2}$$

The load factor against instability of the tower at the load W would thus be, from (1) and (2),

$$\frac{P_E}{W} = 8. \tag{3}$$

These figures of course relate to an idealised uniform tower, but they indicate the practicability of highly flexible towers for suspension bridges.

13.3. The foregoing argument depends, of course, on the ability of the cables to ensure that the column behaves effectively as a column hinged at its top. To discuss this problem, let us first estimate the lateral stiffness provided by the cables at the top of a tower. To simplify this, we will examine the case illustrated in Fig. 34, which

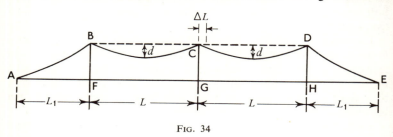

FIG. 34

shows a tower between two equal span bridges. If the back stays AB and DF are nearly straight, the tower tops B and D will be much more stiffly stayed than that at C. We can therefore reasonably discuss a small horizontal displacement ΔL of C relative to B and D as though the latter were fixed. If we move C through ΔL as shown, the dip d of cable BC will decrease and that of CD increase by equal amounts Δd. Equation (39) of Chapter II, taking $L/d = 10$, gives for this change of dip

$$\Delta d \doteqdot 2\Delta L. \tag{4}$$

Thence, from equation (40) in the same chapter, for each cable, the change ΔH in the horizontal component of the cable pull will be

$$\Delta H = \frac{2H}{d} \cdot \Delta L, \tag{5}$$

and the resistance to the movement ΔL will, because of the same change in each span, be equal to twice this value. If each span carries a total loading of w per unit length,

$$H = \frac{wL^2}{8d}. \tag{6}$$

Inserting this in (5), we thus have for the total resistance ΔR to the displacement ΔL,

$$\Delta R = 2\Delta H = \frac{w}{2}\left(\frac{L}{d}\right)^2 \Delta L. \qquad (7)$$

Inserting our value $\frac{L}{d} = 10$, we thus have for the stiffness provided by the cables

$$\frac{\Delta R}{\Delta L} = 50w. \qquad (8)$$

We have next to consider the stiffness required at a tower top just to convert its first mode of instability under end load from the form (*a*) to the form (*b*) in Fig. 35.

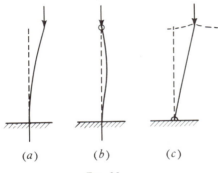

(*a*)   (*b*)   (*c*)

Fɪɢ. 35

An extreme case of this kind arises when the tower is hinged at its base as has been adopted in some suspension bridges of moderate span. In this case, the cable stiffness (8) has to be adequate to prevent the tower rotating about its base in the manner indicated in Fig. 35(*c*). The critical value for the end load *P* for instability in this mode* is

$$P = \beta h, \qquad (9)$$

where *h* is the height of the tower and β is the stiffness of the lateral restraint at the top of the tower. Now the load *P* on the tower will, for a tower with similar spans *L* on each side, be *wL*, where *w* is the total loading (assumed uniform) per unit length along the bridge spans. The restraint will be adequate to prevent this instability, therefore, provided

$$\beta > \frac{P}{h} > w \cdot \frac{L}{h}. \qquad (10)$$

* See p. 78 of Ref. 70 (1st ed.).

Now h will always be greater than d, the dip of the cable. Taking $h/d = 2$ as an upper limit, and $d = L/10$, (10) indicates that in practice the critical stiffness will be at least

$$\beta = 5w. \tag{11}$$

Comparing this with the actual stiffness of $50w$ from (8) provided by the cables in the same circumstances, we see that in practice the cable will usually easily prevent the development of column instabilities of the forms shown in Figs. 35(*a*) and (*c*).

13.4. Having established the stability of the flexible tower, we can outline the necessary calculations for the study of its strength, apart from wind loading. Two loading cases arise:

   (1) All spans fully loaded, giving a peak compression load P_1 on the tower, accompanied by a horizontal displacement of the top of the tower Δ_1.

   (2) Central and one side span fully loaded, giving a compression load of P_2 on the tower with a peak displacement Δ_2 of the tower top.

In each of these cases, the horizontal displacement Δ will be that due to the loads on the bridge spans plus the effects of temperature, chosen to be the most adverse for the conditions specified in (1) or (2). To a good approximation, Δ can be calculated on the assumption that the towers offer no resistance to spanwise movement.

We can now proceed, for a given loading case, to estimate the stresses in the tower, as follows:

   (*a*) Assume a tower deflection form to correspond with the tower top deflection Δ.

   (*b*) With this form, calculate the out-of-balance spanwise force F at the top of the tower necessary, with the end load P, to produce Δ.

   (*c*) With the value of P and F thus determined, calculate the deflection of the tower everywhere up its height.

   (*d*) With the new deflection form resulting from (*c*), repeat the calculations (*b*) and (*c*), and so on until the starting and finishing forms are sufficiently close.

   (*e*) Increase Δ by an amount δ chosen to cover any built-in imperfection in the tower structure, such as initial curvature, accidental offset of the load P, or variations from specification in the thicknesses of the component parts of the tower as fabricated. In the towers of a modern long span bridge, it may well be necessary to adopt a value for δ of several inches.

   (*f*) With the results of (*d*) and (*e*), calculate the corresponding compression stresses in the tower, treated as a cantilever, including wind effects.

The peak stresses deduced in this way have then to be compared with the yield stresses of the steel, reduced if necessary by an appropriate allowance for residual stresses due to the processes of fabrication (rolling of plates or stiffeners, welding, etc.).

13.5. This procedure involves the assumption that the design of the tower cross-section (often cellular) is such that any tendency for local buckling of its component parts under the heavy compression stresses has been prevented, at least at stresses below the yield stress of the steel.

Now the towers of most modern bridges are built up of external plates, riveted or welded together, and stiffened by a system of internal longitudinal and transverse members and plates. A variety of possible modes of local buckling have thus to be investigated. For a tower with a stiffening structure of the type illustrated in Fig. 36, for example, it will be necessary to study, separately in the

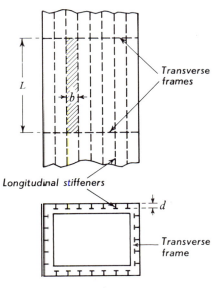

FIG. 36

first instance, and later jointly so far as practicable, the following modes of elastic buckling:

(*a*) Buckling of the component plate shown shaded, of length L and width b, treating its boundaries as simply supported.

(*b*) Buckling of a longitudinal stiffener, of length L and outstanding width d, treating its junction at the outer plating as simply supported.

(*c*) Buckling of the whole side of the box, with its boundaries at the corners and at the diaphragms treated as simply supported, allowing for the integral action of plate and stiffeners.

In addition to such studies, it will be necessary to investigate whether the transverse diaphragms or frames at the pitch *L* are adequate to maintain the cross-sectional shape of the tower and to impose nodes at their positions upon any modes of local buckling of the tower plating and stiffeners.

All such investigations commonly involve analysis by energy methods such as have been developed by aeronautical engineers for wing and fuselage design. For a general text and useful references to detailed literature of this kind, reference may be made to G. Gerard's book [71] on structural stability and to a paper by J. C. Chapman and J. E. Slatford.[72]

13.6. We have concentrated on some design problems affecting the "legs" of the towers and made no reference to the cross-bracing between the legs, whether of triangulated form as in the Forth bridge or portal form as in the Severn bridge.

The bracing is of course brought into action to resist lateral wind forces, both direct on the tower legs and indirect at the tower top and at the deck level from the cables and the deck structure. No special problems arise here and it will suffice to note that, as would be expected, the triangulated form is the more efficient; the portal form, when adopted, has usually been chosen to please architectural opinion.

CHAPTER 14

Fatigue in Suspension Bridges

14.1. It may seem somewhat pessimistic to discuss fatigue in relation to suspension bridges, if only because no large scale failures are known to have occurred due to this cause. Moreover, in general it can fairly be said that in many parts of such bridges the main loads are due to the dead weight of the bridge and therefore not subject to much fluctuation. But the same "laissez faire" attitude was at one time taken in relation to the aerodynamic stability of suspension bridges, and, as we have seen, the resulting inaction ended with the Tacoma disaster.

14.2. There are in fact a number of developments in the design and use of suspension bridges that appear to the author to justify giving the matter some attention here, even though this book is mainly devoted to theory. Firstly, there is the very proper and natural trend for such bridges, particularly the very large ones, to be more and more "tailored" to their purpose and with increasing efficiency. Secondly, again very properly and naturally, designers are turning more and more to steels and other metals that are stronger in a static sense than those hitherto used, but by no means correspondingly stronger in their resistance to fatigue. Thirdly, the vehicular traffic on suspension bridges, in common with many others, is rapidly increasing in frequency and, with goods vehicles, is still increasing in local intensity when measured by axle loads. Finally, economic pressures still put large suspension bridges into the "monumental" class* of structure that is expected to last for a hundred years or more.

All four of these considerations make the liability of a suspension bridge to fatigue trouble greater than hitherto, and parallel, to a remarkable degree, the situation that arose with aeroplanes just before and during the last war, and that led to minor and major fatigue disasters culminating in the Comet tragedies.†

14.3. Before we consider in turn the main components of a suspension bridge from this fatigue standpoint, it is well to note first that nearly all large suspension bridges are, and will commonly be, built over tidal estuaries. They are built in such situations very often to replace overloaded and outmoded ferry boats. This automatically

---

* cf. page 127 of Ref. 73.
† Ref. 73, pages 147 and 148.

puts such a bridge over salt water, often in an exposed and windy site. As a result many suspension bridges are, and will be, forced to exist in a marine atmosphere. One has only to walk across one such bridge on a windy day to taste the salty conditions often surrounding the bridge structure.

We thus see that not only must these bridge structures be well protected against corrosion—and this is fairly easy though expensive—but the fatigue properties under such corrosive conditions must also be studied. And it is well known, not only that fatigue is accelerated in marine atmospheres, but also that complete protection against such acceleration is largely impracticable.

14.4. Cables and Hangers

In the light of these general comments let us now consider the fatigue problems that are most likely to arise in suspension bridges, taking the main components in turn, and trying to draw attention to the research, design and inspection points specially deserving attention.

It is natural first to consider the main cables. These are commonly of high tensile steel wire that well illustrate the second point made in paragraph 14.2: their fatigue properties are low compared with their specially enhanced strength. This consideration applies also to the special "connectors" used to join one length of wire to another during the continuous "reeling" process commonly used in erection; these have a static strength at least equal to the uncut wire, but their fatigue performance is poorer.[74] However, the main cables are loaded mainly by the weight of the bridge itself and the fluctuations due to traffic are, for long span bridges, not large. In these circumstances, one suspects that special local circumstances will alone be found to produce fatigue problems. Two such circumstances may have some importance: firstly, the occurrence of "pitting" in the cable wires due to corrosion in any region of damage to the protecting sheathing and painting; and secondly, the development of "fretting" between adjoining wires, due, for example, to changing curvature of the cable at the tower tops (arising from movements in the vertical or horizontal planes), or at the points of attachment of the hangers causing small relative longitudinal movements of the individual wires or strands. Both these possibilities have received some attention at Bristol, the first in connection with some corrosion of the main cables of the Newport Transporter Bridge, and the second following upon T. A. Wyatt's paper[75] on the effects of curvature changes and a knowledge of the great importance attached nowadays by fatigue specialists to the initiation and aggravation of fatigue by fretting.

A further source of such fretting may sometimes arise due to the fracture of wires during the erection process. It is well known that a small percentage of broken wires do exist in most main cables; and at such discontinuities, particularly when in wires near the cable surface and where the protective soft iron binding wires have lost, by creep or otherwise, part of their initial tension, some fretting due to relative longitudinal movements may well occur.

14.5. If we turn next to the hangers, in smaller and earlier suspension bridges these were commonly not cables but wrought iron rods hinged at their ends. With a vertical system of such rods it is clear that, if they are not hinged at the ends to allow free movement in the planes of the main cables, local bending will occur at the rod ends whenever the main cables and bridge deck move relatively to each other in a longitudinal direction. This of course happens in most modes of oscillation of the bridge and is particularly severe upon the shorter hangers near mid-span. This action was the cause of hanger failures in a number of early small span bridges and taught engineers to hinge the hanger rods properly and ensure that they are free to "work".*

But we are now more usually dealing with hangers made of cables, and, in many cases, either no hinges are provided at the ends or the pins there are so large in diameter as to produce a large frictional resistance to any rotation. As a result, reliance is placed, deliberately or otherwise, upon the flexibility of the cable, particularly near its ends.

A further source of aggravation of the conditions in these regions arises from any tendency for the hanger cables to "sing" in the wind. Aeolian and other type lateral oscillations can be produced by a lateral wind, depending mainly on the surface and diameter of the cables and the wind speed, that give rise to both steady and travelling waves in the hangers causing frequent local bending at the hanger ends.

To add further to the severity of the conditions in these regions, it is clear, because of the necessary jointing processes used in construction, as well as in subsequent use, that these regions, particularly at the bottom ends, will be most liable to local damage (during manufacture and in use), to the presence of "foreign bodies", and to corrosion. It has also to be noted that the hangers are subjected to relatively greater fluctuations of tension than the main cables.

It is not intended to exaggerate the possibilities of fatigue here, but it seems to the author that, in modern suspension bridges, the hangers, whether vertical or inclined, are the part of the cable system

---

* Thus it is still the custom to keep the hinges at the ends of the shorter hangers on the Clifton Suspension Bridge well oiled; nevertheless, frequent working over one hundred years has clearly ovalled the holes through which the pins pass.

that is most vulnerable to fatigue, particularly at their lower ends. It is for this reason that some of the fatigue research work at Bristol is directed to conditions relevant to such hangers. For this reason also, periodic inspection of the lower ends of hangers is recommended.

14.6. The foregoing discussion on the possibilities of fatigue in the cables and hangers of a suspension bridge points to the special importance of local bending at the ends of such members. An examination of this bending in general terms is possible if we assume that the separate wires of a cable are held so securely together that no relative motion, transverse or longitudinal, is possible. In this case it is relevant to examine a cable as a solid thin rod of diameter d and length l between ends that are both fixed in space and in direction, and subjected to a uniform lateral (wind) loading w per unit length when the tension in the cable is T, giving rise to an average stress f.

Under these simplified conditions, the bending moment acting on the cable at each end is

$$M_1 = wj^2 \left(\frac{\frac{1}{2}\theta - \tanh \frac{1}{2}\theta}{\tanh \frac{1}{2}\theta} \right), \tag{1}$$

and the central deflection due to w is

$$y_0 = \frac{wj^2}{8T} \left\{ \frac{4\theta(1 - \cosh \frac{1}{2}\theta)}{\sinh \frac{1}{2}\theta} + \theta^2 \right\}, \tag{2}$$

where $j = \sqrt{\dfrac{EI}{T}}$ and $\theta = l/j$*.

Now for the cables used in suspension bridges, j is very small compared with l, so that $\theta \to \infty$. Thus, from (1),

$$M_1 \to \frac{wl}{2} \sqrt{\frac{EI}{T}}, \tag{3}$$

and from (2),

$$y_0 \to \frac{wl^2}{8T}. \tag{4}$$

Taking these values of M_1 and y_0, and treating the rod as a solid one with a second moment of area equal to $\dfrac{\pi}{64}d^4$, it is easy to show from (3) and (4) that, if p is the bending stress due to M_1, then

$$\frac{p}{f} = 8 \cdot \frac{y_0}{l} \cdot \sqrt{\frac{E}{f}}. \tag{5}$$

* See Ref. 76.

For a steel rod with $E = 13,400 \text{ ton/in}^2$ and $f = 40 \text{ ton/in}^2$ (i.e. about one third of the breaking strength of suspension bridge wire), equation (5) gives

$$\frac{p}{f} = 146 \cdot \frac{y_0}{l}. \tag{6}$$

Thus the end bending stress (p) will be about one quarter of the average tension stress (f) when $y_0/l = 1/600$, and become equal to one half f when $y_0/l = 1/300$. For the various reasons mentioned in paragraphs 14.4 and 14.5, such deflections will certainly sometimes occur in some of the cables and hangers of a suspension bridge, though the corresponding bending stresses may well be lower than in the case of a solid rod.

14.7. Deck Structures

The deck structure of a suspension bridge must broadly present the fatigue problems—or the lack of them—of many steel girder bridges, and on that account a general consideration of their fatigue conditions and characteristics is inappropriate here. Attention is therefore drawn to only two special considerations, both of which arise from the special efforts that are naturally made by suspension bridge designers to minimise deck weights.

The first concerns the very general adoption of jointing by welding on a scale and of a kind that is only matched by that adopted in shipbuilding. In both cases the structures are large and necessarily fabricated in big units for subsequent assembly in the complete structure; in both cases, at least if we think of naval ships, steels of high strength needing increased care in welding, and in welding control and inspection, are used. Admittedly ship structures are subjected to specially severe load fluctuations, but nevertheless it would be strange if, in the course of time, inspection did not find fatigue cracks developing in the deck structures of some suspension bridges. It is so well known in other fields that welding can easily result in small initial cracks. either actual or effective as at inclusions, that careful and regular inspection of these structures seems specially desirable.

Inspection is emphasised here because the author believes that, as in aeronautics, there is a very good chance in welded bridge structures of finding fatigue cracks before they have reached dangerous or "critical" lengths.

The second point to be made here relates to the actual deck surface. Here it is becoming customary for the roadway surface to consist of some kind of asphalt an inch or two thick placed directly on steel sheeting stiffened by longitudinal "stringers" spaced some eighteen inches apart and spanning between heavier transverse members or frames. As a result, in the road case, concentrated wheel loads tend to flex the thin sheet between and over the stringers; and such actions occur with the passage of every wheel. Moreover, the "lane" system of traffic control ensures not only that the heavier traffic shall concentrate in the slow lane, but also that the near wheels shall all pass over almost the same deck panels.

It is clear that there may well be a fatigue problem here, particularly in the region of the welded joints in the plating. A natural design consideration could be to examine the value of using stiffer plates, or intermediate stringers, in the regions specially concerned.

14.8. Nothing will here be said of fatigue in suspension bridge towers; because of the prevailing compression therein, fatigue problems may well not arise in practice. However, there is one detailed feature of modern bridge construction that merits attention here. This is the increasing use of highly prestressed bolts. In particular, reference may be made to the bolts used for securing the clamps by which the hanger loads are transmitted to the main cables.

Such bolts are often of very high strength steel and during assembly tensioned in position to their yield stress to ensure a known high initial tension. In many cases this stress is not altered much by the live loading—due to traffic or wind—on the bridge, and the joints can be so designed as to avoid fluctuations of shear load also. But in some designs this is not so and in such cases stress fluctuations may have severe effects.

14.9. It seems desirable, in conclusion, to make some reference to the two current philosophies of design against fatigue—"safe life" and "fail safe". As is now usual in the aircraft world where these terms grew up, both philosophies seem to be relevant to suspension bridge design. All those members which are constructed of cables comprising large numbers of discrete wires are likely, should they suffer from fatigue, to show it by the premature failure of individual wires. The tension load has many possible paths in a cable, which is by its nature "fail-safe", and fortunately so.

But the deck structure of a suspension bridge is in a more intermediate category. It should clearly be made as "fail-safe" as possible by the use of materials and design details likely to reduce the probability of fatigue cracks, and to ensure that any such grow only slowly and are not able to become dangerously large. On the other hand, the aim should also surely be to give the structure as a whole a "safe-life" well in excess of its probable life in practice.

The degree of redundancy in the main components of a suspension bridge is perhaps least in the towers and the anchorages. The towers are largely in compression and unlikely to suffer from fatigue in a widespread form, but the anchorages include some highly localised structures in tension. At such points, almost more than in the deck, a "safe-life" policy seems essential, whether or not some quality of "fail-safe" can also be introduced.

14.10. In this chapter the author has aimed to draw attention to the possible fatigue problems special to suspension bridges rather than to discuss their detailed solution. Such discussions would clearly be more appropriate elsewhere and many general references are available.* In so far as these problems arise mainly from the way in which suspension bridge structures function and are loaded, it is hoped that the matter of this chapter is not only opportune but also not inappropriate.

---

* See, for example, Ref. 77.

Select Bibliography

THE following books, in English and readily accessible, are selected as including interesting statements of suspension bridge theory.

1. *A Practical Treatise on Suspension Bridges*, by D. B. Steinman, 2nd edition, John Wiley, 1929.
2. *The Theory and Practice of Modern Framed Structures*, Part II, by J. B. Johnson, C. W. Bryan and F. E. Turneaure, 9th edition, John Wiley, 1911.
3. *The Collected Papers of S. P. Timoshenko*, McGraw-Hill, 1953.
4. *Design of Suspension Bridges*, by Arne Selberg, Trondheim, 1946.
5. *Aerodynamic Stability of Suspension Bridges*, by F. B. Farquharson, Univ. of Washington Press, U.S.A., 1949.
6. *Dynamic Instability*, by Y. Rocard (translation from the French by M. L. Meyer), Crosby Lockwood & Son, 1957.

In addition, for a very extensive bibliography of suspension bridges, including a section on works on the theory, reference is made to:

7. *A History of Suspension Bridges in Bibliographical Form*, by A. A. Jakkula, Bulletin of the Agricultural and Mechanical College of Texas, U.S.A., 1941.

References

1. F. W. Robins, *The Story of the Bridge*, Cornish Bros., 1948.
2. J. Dredge, Mathematical demonstration of the principles of Dredge's patent iron bridges, *Mechanics Magazine*, 1843.
3. P. W. Barlow, Combining girders and suspension chains, *J. Franklin Inst.*, 1858.
4. J. A. Roebling, Passage of the first locomotive over the suspension bridge over the Falls of Niagara, *J. Franklin Inst.*, 1855.
5. D. B. Steinman and S. R. Watson, *Bridges and Their Builders*, 1941, p. 219.
6. C. Clericetti, The theory of modern American suspension bridges, *Proc. Inst. Civ. Engrs*, Vol. 60, 1880.
7. C. B. Bender, Suspension bridges of any desired degree of stiffness, *Van Nostrand's Engineering Magazine*, 1881.
8. M. Levy, Mémoires sur le calcul des ponts suspendus rigides, *Ann. Ponts Chaussées*, 1886.
9. A. Castigliano, *Théorème de l'Équilibre des Systèmes élastiques et ses applications*, Paris, 1879. Translated by E. S. Andrews and published as *Elastic Stresses in Structures* in 1919.
10. J. Melan, Theorie der eisernen Bogenbrücken und der Hängebrücken, *Handbuch der Ingenieurwissenschaften*, Leipzig, 2nd ed. 1888; 3rd ed. 1906.
11. D. B. Steinman, *A Practical Treatise on Suspension Bridges*, John Wiley, 1st ed. 1922; 2nd ed. 1929.
12. The statics of bridges—the suspension chain, *Civil Engineer and Architects' Journal*, 1862.
13. J. Melan, (trans. by D. B. Steinman), *Theory of Arches and Suspension Bridges*, Chicago, 1913.
14. S. Timoshenko, Steifigkeit von Hängebrücken, *Z. angew. Math. Mech.*, 1928; The stiffness of suspension bridges, *Proc. Amer. Soc. Civ. Engrs*, 1928; The stiffness of suspension bridges, *Trans. Amer. Soc. Civ. Engrs*, Vol. 94, 1930.
15. R. J. Atkinson and R. V. Southwell, On the problem of stiffened suspension bridges and its treatment by relaxation methods, *J. Inst. Civ. Engrs*, Vol. 11, 1939.
16. C. D. Crosthwaite, The corrected theory of the stiffened suspension bridge, *J. Inst. Civ. Engrs*, Vol. 27, 1946.
17. H. Bleich, *Die Berechnung verankerter Hängebrücken*, Julius Springer, Vienna, 1935.
18. A. G. Pugsley, A flexibility coefficient approach to suspension bridge theory, *J. Inst. Civ. Engrs*, Vol. 32, 1949.
19. G. B. Airy, On the use of suspension bridges with stiffened roadways, *Proc. Inst. Civ. Engrs*, Vol. 26, 1867.
20. S. Hardesty and H. E. Wessman, The preliminary design of suspension bridges, *Proc. Amer. Soc. Civ. Engrs*, Vol. 65, 1939.

21. A. G. Pugsley, The gravity stiffness of a suspension bridge cable, *Quart. J. Mech. Appl. Math.*, Vol. 5, 1952.
22. A. G. Pugsley, A simple theory of suspension bridges, *J. Inst. Struct. Engrs*, Vol. 31, 1953.
23. A. G. Pugsley, Note on the foundation analogy for the approximate analysis of suspension bridges, *J. Inst. Struct. Engrs*, Vol. 40, 1962.
24. C. F. P. Bowen and T. M. Charlton, A note on the approximate analysis of suspension bridges, *J. Inst. Struct. Engrs*, Vol. 45, 1967.
25. *The Failure of the Tacoma Narrows Bridge*, Report of Board of Engineers, 1941.
26. R. A. Frazer and C. Scruton, *A Summarised Account of the Severn Bridge Aerodynamic Investigation*, H.M.S.O., 1952.
27. F. Bleich, Dynamic instability of truss-stiffened suspension bridges under wind action, *Trans. Amer. Soc. Civ. Engrs*, 1949.
28. A. G. Pugsley, Some experimental work on model suspension bridges, *J. Inst. Struct. Engrs*, Vol. 27, 1949.
29. D. B. Steinman, Rigidity and aerodynamic stability of suspension bridges, *Trans. Amer. Soc. Civ. Engrs*, 1945.
30. D. E. Walshe, A résumé of the aerodynamic investigations for the Forth Road and the Severn Bridges, *Proc. Inst. Civ. Engrs*, Vol. 37, 1967.
31. D. Gilbert, *Phil. Trans. Roy. Soc.*, Part III, 1826.
32. E. J. Routh, *Treatise on Analytical Statics*, Vol. I, 1st ed., 1891.
33. T. J. Poskitt, The application of elastic catenary functions to analysis of suspended cable structures, *J. Inst. Struct. Engrs*, Vol. 41, 1963.
34. W. J. M. Rankine, *A Manual of Applied Mechanics*, 1st ed., 1858.
35. *Idem, ibid.*, Section 2, Chapter II.
36. H. Straub, *A History of Civil Engineering*, Leonard Hill, 1952, Chapter VI.
37. A. J. S. Pippard and J. F. Baker, *The Analysis of Engineering Structures*, Edward Arnold, 1st ed., 1936; 3rd ed., 1957.
38. J. B. Johnson, C. W. Bryan and F. F. Turneaure, *Theory and Practice of Modern Framed Structures*, Part II, 9th ed., 1910.
39. D. M. Brotton, N. W. Williamson and M. Millar, *J. Inst. Struct. Engrs*, Vol. 41, 1963.
40. M. T. Godard, Recherches sur le calcul de la résistance des tabliers des ponts suspendus, *Ann. Ponts Chaussées*, 1894.
41. F. Bleich, *The Mathematical Theory of Vibration in Suspension Bridges*, U.S. Department of Commerce, Washington, 1950.
42. T. J. Poskitt, Structural analysis of suspension bridges, *Proc. Amer. Soc. Civ. Engrs*, 1966.
43. C. F. P. Bowen and T. M. Charlton, A note on the Approximate analysis of suspension bridges, *J. Inst. Struct. Engrs*, July 1967.
44. R. V. Southwell, *Relaxation Methods in Engineering Science*, Oxford, 1940.
45. S. Timoshenko, Theory of suspension bridges, *J. Franklin Inst.*, Vol. 235, 1943.
46. C. D. Crosthwaite, Analysis of the long span suspension bridge, *3rd Congress (Liège), International Association for Bridge and Structural Engineering*, Zürich, 1948.

47. M. HETENYI, *Beams on Elastic Foundations*, Univ. of Michigan Press, 1946.
48. A. R. FLINT and A. G. PUGSLEY, Some experiments on Clifton Suspension bridge, *Conference on the Correlation between Calculated and Observed Stresses and Displacements in Structures*, Inst. Civ. Engrs, September 1955.
49. J. H. ROHR, On the oscillations of a suspension chain, *Trans. Cam. Phil. Soc.*, Vol. 9, 1851.
50. E. J. ROUTH, *Advanced Rigid Dynamics*, 6th ed., Macmillan, 1905.
51. A. G. PUGSLEY, On the natural frequencies of suspension chains, *Quart. J. Mech. Appl. Math.*, Vol. 2, 1949.
52. D. S. SAXON and A. S. CAHN, Modes of Vibration of a suspended chain, *Quart J. Mech. Appl. Math.*, Vol. 6, 1953.
53. N. K. CHAUDHURY and D. M. BROTTON, Analysis of vertical flexural oscillations of suspension bridges by digital computer, *Symposium on Suspension Bridges*, Lisbon, 1966.
54. A. SELBERG, Oscillation and aerodynamic stability of suspension bridges, *Acta Polytechnica Scandinavica*, Trondheim, 1961.
55. Y. ROCARD, *Dynamic Stability*, translated from the French by M. L. Meyer, Crosby Lockwood, London, 1957.
56. F. B. FARQUHARSON, *Aerodynamic Stability of Suspension Bridges*, Bulletin 116, University of Washington, 1949.
57. *Wind Effects on Buildings and Structures*, Report on the Symposium at the National Physical Laboratory, H.M.S.O., 1965.
58. D. B. STEINMAN, *Suspension Bridges and Cantilevers*, D. Van Nostrand, 1913.
59. I. S. MOISSEIFF and F. LIENHARD, Suspension bridges under the action of lateral forces, *Proc. Amer. Soc. Civ. Engrs*, 1932.
60. A. SELBERG, Calculation of lateral truss in suspension bridges, *International Association for Bridge and Structural Engineering*, Vol. 7, Zürich, 1944.
61. A. G. DAVENPORT, The action of wind on suspension bridges, *International Symposium on Suspension Bridges*, Lisbon, 1966.
62. N. A. V. PIERCY, *Aerodynamics*, E.U.P., 1947.
63. H. G. KUSSNER, Schwingungen von Flugzeugflügeln, *Luftfahrtforschung*, Vol. 4, 1929.
64. E. G. BROADBENT, *The Elementary Theory of Aeroelasticity*, Bunhill Publications, 1954.
65. C. SCRUTON, An experimental investigation of the aerodynamic stability of suspension bridges, *3rd Congress (Liège)*, *International Association for Bridge and Structural Engineering*, Zürich, 1948.
66. A. SELBERG, Dampening effects in suspension bridges, *International Association for Bridge and Structural Engineering*, Vol. 10, Zürich, 1950.
67. R. A. FRAZER and W. J. DUNCAN, On the criteria for the stability of small motions, *Proc. Roy. Soc.*, Series A, Vol. 124, 1929.
68. A. G. PUGSLEY, A simplified theory of wing flutter, *A.R.C. R. & M.* 1839, 1937.
69. A. SELBERG, Aerodynamic effects on suspension bridges, Report on

the Symposium at the National Physical Laboratory on *Wind Effects on Buildings and Structures*, Vol. II, H.M.S.O., 1965.
70. S. TIMOSHENKO, *Theory of Elastic Stability*, McGraw-Hill, 1st ed., 1936; 2nd ed., 1961.
71. G. GERARD, *Introduction to Structural Stability*, McGraw-Hill, 1962.
72. J. C. CHAPMAN and J. E. SLATFORD, Design of stiffened plating in compression, *The Engineer*, 1959.
73. A. G. PUGSLEY, *Safety of Structures*, Arnold, 1966.
74. M. S. G. CULLIMORE, The fatigue behaviour of sleeved and swaged joints in 0·2 in. diameter high tensile steel wire, *J. Inst. Struct. Engrs*, Vol. 44, 1966.
75. T. A. WYATT, Secondary stresses in parallel wire suspension cables, *Trans. Amer. Soc. Civ. Engrs*, Vol. 128, 1963.
76. R. J. ROARK, *Formulas for Stress and Strain*, 4th ed., McGraw-Hill, 1938.
77. R. B. HEYWOOD, *Designing against Fatigue*, Chapman & Hall, 1962.